I0467988

Analyzing Legacy U.S. Geological Survey Geochemical Databases Using GIS— Applications for a National Mineral Resource Assessment

Douglas B. Yager, Albert H. Hofstra, and Matthew Granitto

Techniques and Methods 11–C5

U.S. Department of the Interior
U.S. Geological Survey

U.S. Department of the Interior
KEN SALAZAR, Secretary

U.S. Geological Survey
Marcia K. McNutt, Director

U.S. Geological Survey, Reston, Virginia: 2012

For more information on the USGS—the Federal source for science about the Earth, its natural and living resources, natural hazards, and the environment, visit http://www.usgs.gov or call 1–888–ASK–USGS.

For an overview of USGS information products, including maps, imagery, and publications, visit http://www.usgs.gov/pubprod

To order this and other USGS information products, visit http://store.usgs.gov

Suggested citation:
Yager, D.B., Hofstra, A.H., and Granitto, Matthew, 2012, Analyzing legacy U.S. Geological Survey geochemical databases using GIS—Applications for a national mineral resource assessment: U.S. Geological Survey Techniques and Methods 11–C5, 28 p.

Contents

Figures

Tables

Analyzing Legacy U.S. Geological Survey Geochemical Databases Using GIS—Applications for a National Mineral Resource Assessment

By Douglas B. Yager, Albert H. Hofstra, and Matthew Granitto

Abstract

Legacy geochemical and locality data for 414,304 rock, 397,625 soil and stream sediment, and 335,547 water samples acquired over the 100-year history of the U.S. Geological Survey are digitally compiled in the National Geochemical Database. Upcoming mineral environmental and resource assessments of the Nation will utilize some or all of these data to define geochemically anomalous areas. In such assessments it is advantageous to define the watershed from which sediment, soil, and water samples were derived. The goal of this study was to determine the most expedient methods to delineate and display the source catchments and hydrologic contributing areas of geochemically anomalous sediment, soil, and associated rock samples.

A test area in the southern Egan Range in east central Nevada was used because it had been sampled and assessed as part of the National Uranium Resource Evaluation in the 1970s and as part of the Bureau of Land Management's Wilderness Study Area program in the 1980s. The data generated are representative of those present in the National Geochemical Database for the Western United States. In this study, a geographic information system analysis approach was applied to the Egan Range study area to (1) determine useful geographic information system analysis and display methodologies that efficiently highlight geochemically anomalous sample sites, (2) analyze point data having x-y geographic coordinates and geochemical attributes for sediments and rocks in relation to national hydrologic network and watershed boundary datasets, and (3) describe the methods that are most expedient and effective at delineating the source catchments for anomalous samples and sediment provenance. This study utilized a 1:24,000-scale National Hydrography Dataset that provides a hydrologic network and watershed boundaries for the Nation. The National Hydrography Dataset provides a relatively complete and detailed hydrologic network that has associated flow-routing attributes that permit upstream and downstream relations relative to sample sites to be determined that are, in turn, useful in delineating stream sediment provenance.

Fundamental geographic information system analysis of geochemical data at point locations was a useful first step in delineating anomalous sample sites. Thematic maps that used proportional symbols to represent higher element concentrations with larger symbol sizes quickly aid in highlighting geochemical anomalies. The combined use of gradational colors and proportional symbols is also very effective in showing high metal concentrations. Spatial statistics and hot-spot analysis (which calculates Z-score standard deviation for each sample) identified not only high element-concentration data clusters but also areas that represent the geochemical baseline. We found that watershed boundaries from the National Hydrography Dataset are a useful geographic frame of reference for associating anomalous point data. Geographic information system queries can be used to highlight associated watershed boundary areas that contain anomalous point data. This analysis is fundamental to identifying areas for focusing mineral exploration. Tools available in Environmental Systems Research Institute, Inc., ArcGIS that analyze hydrologic network attributes stored in the National Hydrography Dataset that permit upstream and downstream relations relative to a point to be determined are very effective in delineating local provenance of sediment at a specific site. A software plugin for ArcGIS, the "Hydrography Event Management" tools, also proved effective for managing multiple point features over large areas and resulted in determining stream sediment provenance at multiple sites where samples intersect a stream network. Due to the variable positional accuracy of legacy data collected prior to a global positioning system and mapped on small-scale (large area) base maps, adjustment of sample-site data with the aid of a geographic information system may be necessary to accurately determine watershed contributing areas using automated digital techniques. The Hydrography Event Management tools proved effective in automatically adjusting sample point locations based on search-radius criteria that determined the average distance of a sediment sample point relative to a stream.

Introduction

Mineral exploration geologists commonly analyze stream sediment and soil samples collected from watersheds as a geochemical indicator of potential mineral deposits. Stream sediments represent an integrated sample of materials that have been weathered and transported from soils and bedrocks that are exposed upslope. Historically, stream sediments have been used as part of a mineral exploration strategy since the first explorers began searching for precious metals. For example, gold veins were discovered by identifying placer "shows" of gold in fluvial sediments that were then used to help guide prospectors upstream to lode gold deposits in bedrock outcrops. Such a sampling and analysis approach has helped to prioritize and focus mineral exploration in the field when geochemically anomalous areas are identified or, conversely, to eliminate areas from further investigation (Carranza and Hale, 1997; Seoane and Silva, 1999; Brady and others, 2001; Carranza, 2004). Geochemical anomalies in stream sediments are also used to recognize areas that may be sources of environmental contaminants to surface water and groundwater (Church and others, 2008).

This report discusses a geographic information system (GIS) methodology to utilize GIS tools to analyze existing digital elevation, hydrologic, and legacy geochemical data to delineate mineralized watersheds. A goal of this study was to evaluate existing datasets and GIS methodologies that can help delineate mineralized terrain as part of a current U.S. Geological Survey (USGS) mineral assessment. GIS tools and digital datasets are used in these analyses that were unavailable when a previous USGS mineral assessment was completed between 1996 and 1998 (U.S. Geological Survey, 2002).

Data Sources and Methodology

Legacy Field-Based Mapping and Geochemical Sampling

Geochemical analyses of sediments and rocks determined as part of a Bureau of Land Management (BLM) wilderness study by Rowan and others (1984) in the Egan Range, Nev., are stored in the USGS National Geochemical Database (NGDB) (Smith, 1997; Granitto and others, 2005). The NGDB stores 414,304 rock, 397,625 sediment, and 335,547 water geochemical analyses in the United States. National Uranium Resource Evaluation (NURE) rock, sediment, and soil data were also analyzed in this study. The NURE project was initiated in 1973 and completed in 1984.

Sample Location Accuracy

Positional accuracy and grid sampling density varies for sample geochemistry stored in the NGDB, and the location accuracy is project dependent. BLM wilderness and NURE datasets were collected prior to wide use of global positioning system devices in the field. Thus, when point-data locations were digitized and input into a GIS, the possible errors in site locations could have affected digital mapping and analysis results. In the case of the study by Rowan and others (1984), stream sediment, rock, and water sample localities were compiled on 1:24,000-scale base maps in the field. The dataset of Rowan and others (1984) is likely a best-case scenario for having a relatively high sampling density and positional accuracy because the work was completed by geologists who monitored quality control during all phases of investigation. Sample coordinates used by Rowan and others (1984) were determined using a coordinate grid that was placed on a field sheet to estimate latitude and longitude. Average sample spacing for the BLM wilderness data using the Environmental Systems Research Institute, Inc., ArcGIS "Calculate Distance Band From Neighbor Count" was 757 m. The original positional accuracy of the BLM wilderness study area data was ± 12 m. Data were subsequently transferred to 1:62,500-scale maps for publication. The 1:62,500-scale base maps were subsequently georeferenced in GIS. Location errors could have been introduced because field locations were transferred to 1:62,500-scale base maps having a horizontal accuracy of ±73 m and subsequently digitized. For additional details on sampling methodologies see Rowan and others (1984).

NURE sample localities were recorded in the field on 1:24,000- and 1:48,000-scale base maps and transferred to a clean, master map copy at the same scales. Sample localities recorded on the master map copies were subsequently digitized in the office. Average sample spacing for the NURE data determined using the ArcGIS 10.1 spatial statistics tool utility Calculate Distance Band From Neighbor Count is 2,145 m. Any errors in sample locations were a result of positional inaccuracies recorded in the field. No rigorous sample locality error assessment was completed as part of the NURE project, although during data input into the NGDB, location errors were corrected when field notes had sufficient information to verify that corrections were needed. Sediment samples were commonly collected in primary drainages, although when plotted on GIS maps and in context with a digital hydrologic network, the locations seldom plot on a stream. If samples do not plot on a stream, the locations can be accepted as accurate, or the data can be snapped to a stream using GIS tools.

Geologists will often collect sediment from adjacent segments of a bifurcating stream to assess potential geochemical anomalies of adjacent watersheds. A decision to digitally move a point near stream intersections can be based on defined rules in a GIS. For example, points that plot within 30 m of a stream will be snapped to a stream. These types of adjustments to point data can be manually performed using standard GIS editing tools or automated by assigning a snap tolerance distance as part of the linear referencing tool in ArcGIS. Manual repositioning of points in a GIS editing environment provides the most control of snapping results; however, this approach is time consuming. Thus, as part of a national mineral assessment, automated ArcGIS linear referencing tools are an efficient method to snap points to a hydrologic network, especially when multiple point features are involved that cover large areas.

Digital Mapping Methods

Hydrologic Network and Watershed Boundaries

The high-resolution, 1:24,000-scale National Hydrography Dataset (NHD) is used in this analysis and serves as a principal hydrologic and geographic frame of reference for point-geochemical data and geologic coverages. The NHD provides comprehensive, digital geographic data for surface water features of the United States (Simley and Carswell, 2009) and may be downloaded at *http://viewer.nationalmap. gov/viewer/nhd.html?p=nhd.*

The NHD has flow-route attributes that permit upstream and downstream relations to be determined for hydrologic event features (points) that intersect the NHD network. Each route on a NHD network has a unique reach code. A flow-attribute table stores stream-reach information pertaining to the upstream (100 percent to end) and downstream (0 percent to end) for a reach, as well as between reaches flow-direction attributes. The "Utility Network Analyst" in ArcGIS 10.1 can utilize NHD flow-route information to identify upstream source areas that may have transported sediment to a downstream site. This is an important component for stream sediment provenance investigations discussed below. An example of the 1:24,000-scale NHD hydrologic network for the BLM wilderness study area is shown overlain on high-resolution, world imagery from ArcGIS online resources (fig. 1). Although this area is located in a semi-arid climate, receiving between 10 and 20 cm of precipitation per year (PRISM Climate Group, 2002), each watershed and associated tributary shown in figure 1 could be a potential sediment contributing area during storm events or during snowmelt runoff.

This study also utilized the Watershed Boundary Dataset (WBD) that is included as part of the NHD data-download set of hydrologic features. The WBD is managed by the National Resources Conservation Service and U.S. Geological Survey. This study utilized 8- and 12-digit hydrologic-unit-code (HUC) watershed boundaries with an average area of 1,813 km^2 (700 mi^2) and 36 km^2 (14 mi^2) respectively (U.S. Geological Survey and U.S. Department of Agriculture, 2011). HUCs are digital watershed boundaries grouped by size that are defined by a numeric code schema. The 8- and 12-digit HUCs are readily available for the United States. More detailed, higher resolution WBDs are available in selected areas.

Watersheds delineated on field maps (Rowan and others, 1984) were georectified in GIS and geographically compared with WBDs as part of the NHD datasets. In addition, comparisons of field-derived watersheds were compared to watersheds delineated using 30-m-elevation data downloaded from the USGS National Elevation Dataset (NED) (Gesch and others, 2002), and the GIS Weasel (see Viger, 2008). The GIS Weasel can be used to delineate watersheds for small areas with user-defined locations that track all upstream contributing areas above a point location and contributing area threshold values used to define a watershed. GIS Weasel derived watersheds are more detailed than the 12-digit WBDs. For detailed information on using the GIS Weasel, a tutorial is available (see Viger and Leavesley, 2006).

Watersheds defined using the GIS Weasel permit one-plane (whole watersheds) and two-plane (watersheds that are separated into hydrologic right- and left-bank sections) delineations. When two-plane watersheds are defined, the specific part of a watershed that is a potential contributing area to a geochemically anomalous sediment sample can be more accurately determined. A unique identification that signifies the part of a two-plane watershed having high metal concentrations can be geographically associated with a geochemically anomalous sample site using GIS. This would aid field investigations in focusing exploration in the part of a watershed that could be most favorable for mineral exploration. Figure 2 shows examples of one- and two-plane watersheds determined for part of the BLM wilderness study area. Analysis using the GIS Weasel as part of a national mineral resource assessment would likely only be feasible in areas requiring detailed study.

Symbolizing Raw Point Data

BLM wilderness area point-geochemical sample data for soils, sediments, water, and rocks were retrieved from the USGS NGDB archive for the study area. The datasets used for this analysis were collected as part of the NURE investigation and as part of the Rowan and others (1984) BLM wilderness study (table 1). These data are representative, especially for the Western United States, of the types of digital geochemical information available for the Nation and archived in the NGDB.

Legacy geochemical sample data were imported into GIS from databases having point locations with x-y geographic coordinates and associated physical and geochemical attributes. Once imported, x-y point data subsequently could be converted to shape files, coverages, or feature classes in a geodatabase that could then be queried and symbolized in ArcGIS 10.1.

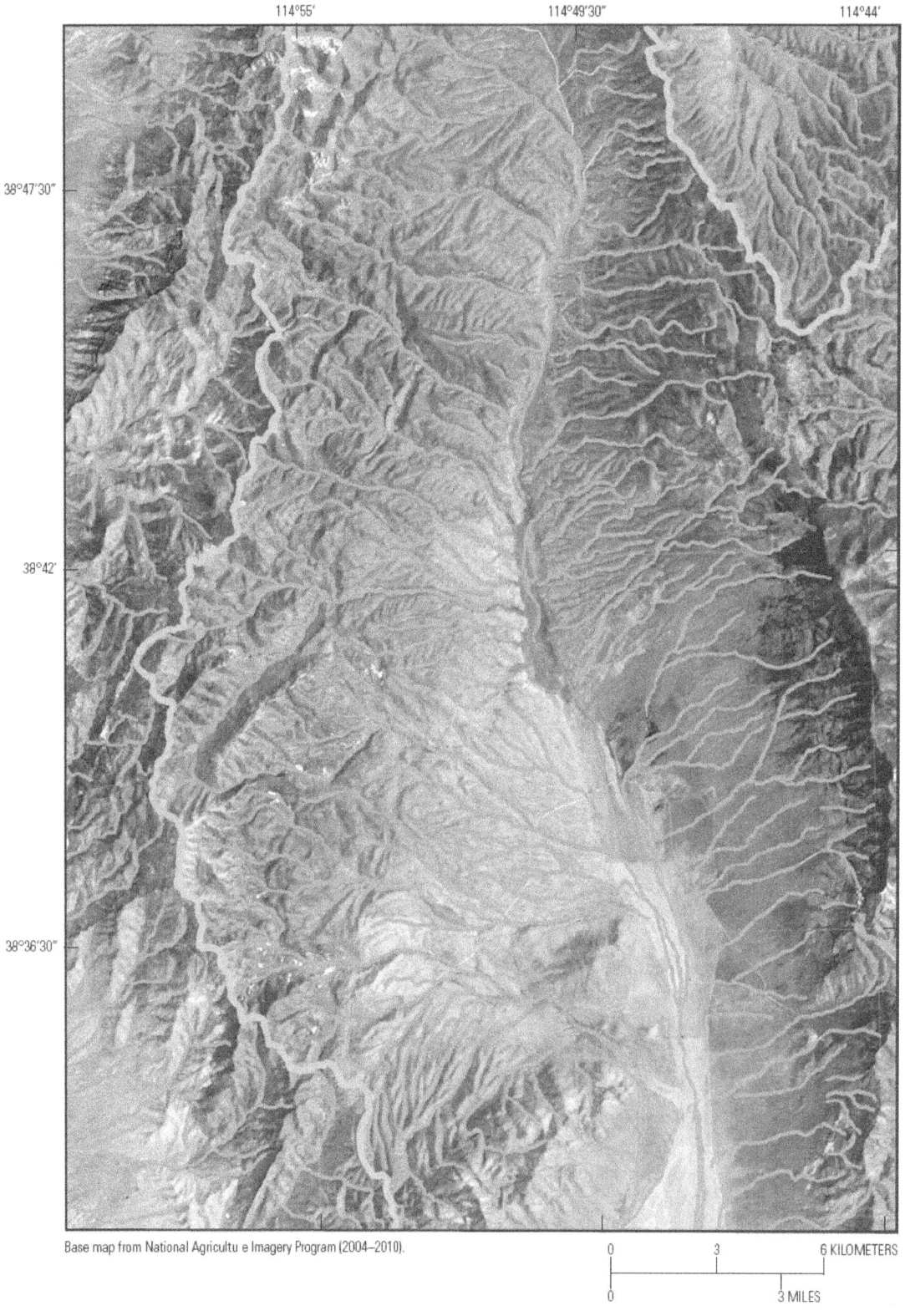

Figure 1.   Image showing 1:24,000-scale National Hydrography Network (dark blue), 8-digit hydrologic-unit-code watershed boundary (light blue), and 1-m resolution National Agricultural Imagery Program base map for part of the Egan Wilderness Study Area.

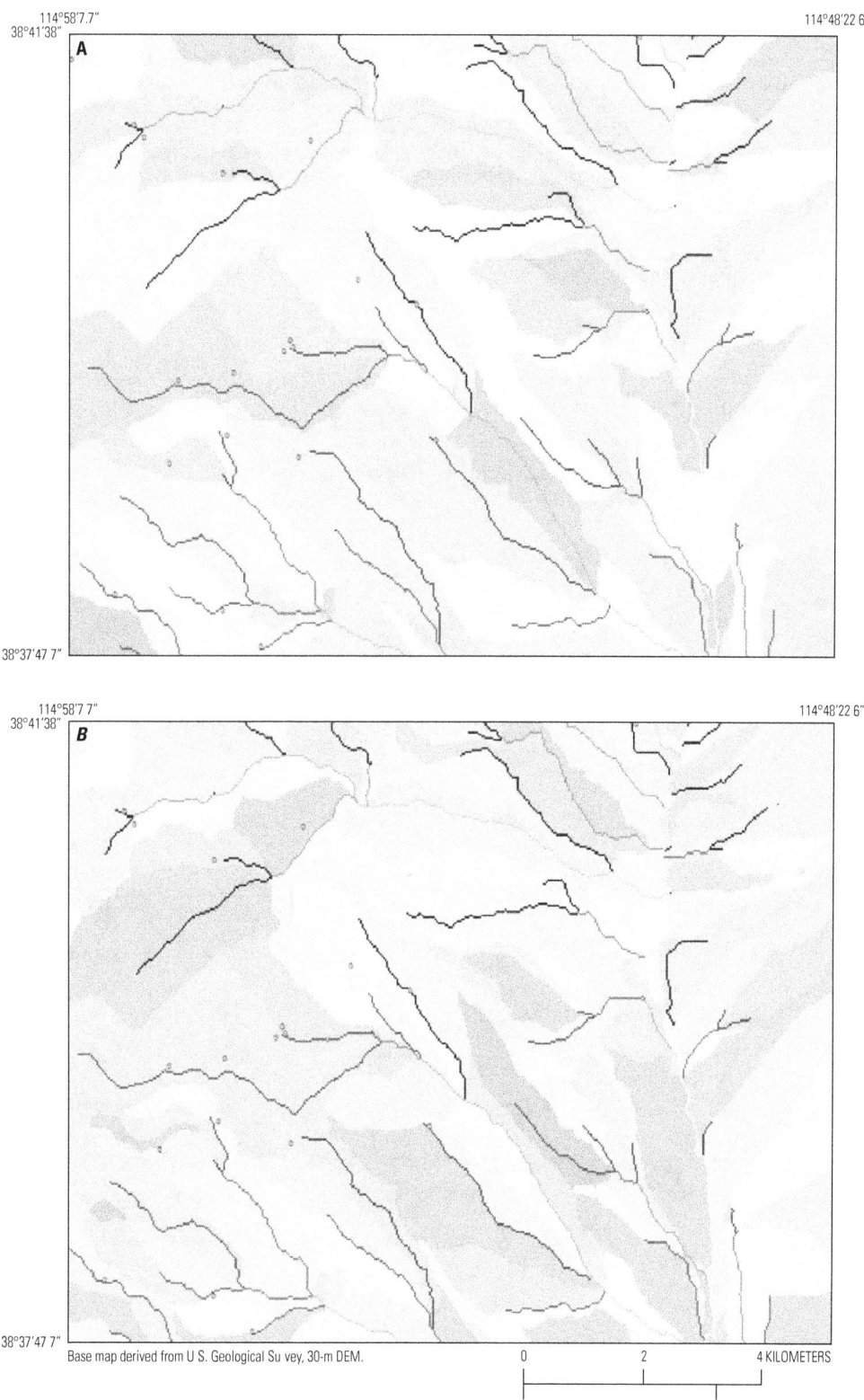

Figure 2.   (*A*) One-plane watersheds, (*B*) Two-plane watersheds and hydrologic network delineated using the geographic information system Weasel software system. Sediment sample locations shown as green dots.

Table 1.   Legacy geochemical data retrieved from the National Geochemical Database and analyzed as part of this study.

[ES, semi-quantitative emission spectrography; AA, atomic absorption spectrometry; alk, alkalinity; IC, ion chromatography; FL, fluorometry; DN, delayed neutron activation counting; NA, instrumental neutron activation; ICP or ICAP, inductively coupled plasma–atomic emission spectrometry; C-forms, forms of carbon; C-S, total carbon and sulfur; C3, nonmagnetic heavy mineral concentrate; CVAA, cold-vapor atomic absorption spectrometry; HGAA, hydride genera-tion–atomic absorption spectrometry; XRF, X-ray fluorescence spectrometry; ISE, ion-specific electrode; Comb, combustion; Rapid Rock, various methods used to determine 10 major oxides and volatiles (CO_2, moisture, bound water, FeO). BLM, Bureau of Land Management; WAS, wilderness study area; NURE, National Uranium Resource Evaluation (NURE); SRL, Savanah River Labs; ORGDP, Oak Ridge Gas Diffusion Plant]

Project	Submitters	Project years	Media	Number of samples	Analyses
BLM WSA, SW:Egan WSA	Rowan, Elisabeth L.	1983	Rock	26	ES, Au-Ag-As-Sb_AA
			Sediment	181	ES, Au-Ag-As-Sb_AA
			C3	169	ES
			Water	1	Metals_AA, IC, alk, SP
NURE	SRL/ORGDP	1980	Sediment	89	ES31, U_FL
			Water	40	ES29, U_DN, NA9, pH, alk
Regional survey and analysis	Grosz, A E., and Grossman, J. N.	2000–2002	Sediment	7	ICP40, Hg_CVAA, As-Se_ HGAA
			Soil	1	ICP40, Hg_CVAA, As-Se_ HGAA
Epithermal Au-Ag	Berger, Byron R.	1981–82	Rock	17	ES, Au-As-Sb-Zn_AA
White Pine, Colo.	Blake, M. Clark, Jr.	1968–69	Mineral	3	K_2O_AA
			Rock	9	ES, RapidRock
Petroleum geology of Paleozoic rocks of Cordilleran miogeosyncline	Poole, Forrest G.	1975	Rock	1	ES, DN, C-forms, Se_XRF
Special test site geologic studies	Sargent, Kenneth A.	1969	Rock	4	ES, RapidRock
Tippecanoe carbonates	Schultz, Leonard G.	1975	Rock	2	ES, DN, C-forms, F_ISE, Majors_XRF, Mg-Na-Li-Rb-Zn-Cd_AA
Tippecanoe carbonates	Schultz, Leonar G.	1975	Rock	3	ES, DN, C-S_Comb, F_ISE, Majors_XRF, Mg-Na-Li-Rb-Zn_AA
W. Interior Tertiary	Claypool, George E.	1975	Rock	5	C-forms
BLM WSA	Nakata, John K.	1986	Mineral	1	K_2O_AA
Geochron	Marvin, Richard F. and McKee, Edwin H.	1968, 84	Mineral	2	K_2O_AA
NURE	SRL	1980	Sediment	89	ICAP31, U_FL
			Water	40	ICAP29, U_DN, NA9, pH, alk

Establishing Geochemical Threshold Values

Determining geochemically anomalous threshold values for geochemical samples, though beyond the scope of this report, is clearly important to consider in mineral and geoenvironmental investigations. In addition, geochemical threshold values for stream sediment samples pose complexities because anomalies may be difficult to determine due to the lack of comprehensive geochemical baseline data. There are multiple statistical and empirical methods to establish anomalous geochemical concentrations. A statistically robust approach is described in Yager and Folger (2003) to map element concentrations that are at, above, or below the mean. This statistical treatment provides an objective dataset to identify geochemical anomalies.

Although no rigorous attempt was done in this study to establish threshold values, this report does show examples of symbology and map query tools available in ArcGIS 10.1 that can be used to highlight anomalies. Natural breaks and standard deviation were used in this study to identify potentially anomalous stream sediment samples. Natural breaks in ArcGIS 10.1 determine those element concentrations that tend to be naturally grouped or clustered, and standard deviation identifies data that are "n" number of standard deviations above or below the mean.

Thematic Mapping of Point-Geochemical Data

A first step in GIS is to analyze the raw point data. Unadjusted point data having x-y and geochemical attribute information were initially imported into ArcGIS 10.1 and thematically mapped and analyzed. An effective way to thematically display geochemical data is to use proportional symbols, with a symbol size increasing as a function of element concentration. Proportional symbol maps for the BLM wilderness and NURE datasets are shown in figures 3 and 4 respectively.

The number of proportional symbol groupings used in this analysis was based on natural breaks. Several classification options are available in ArcGIS 10.1 that allow statistical grouping of geochemical data that include manual, equal, defined, quantile, and geometric intervals, and standard deviation. In addition to proportional symbols, a color scheme was used to portray sites having low concentrations with cool colors (green, blue) and sites having higher concentrations with warmer colors (yellow, orange, red). The combined use of proportional symbols and color to highlight anomalous sites is an effective first step in analyzing legacy geochemical data that quickly and efficiently visually highlight areas of interest for additional analysis.

Thematic Mapping of Watershed Boundary Datasets Using Point-Geochemical Data

In addition to thematically mapping raw point data, delineation of WBDs of various scales that contain anomalous point-geochemical sample data is useful to identify areas for mineral exploration. Points that are selected based on geochemical criteria and subsequently output as a separate shapefile can be used to identify geographically associated 8- and 12-digit WBDs used in this study. Select by location queries in ArcGIS can be used to delineate regional-scale watersheds that contain anomalous geochemical point data. Figure 5 shows the results of a select-by-location query that delineates an 8-digit HUC having a copper (Cu) anomaly. Finer resolution 12-digit WBDs, however, are required to more accurately delineate the watershed areas that may be the most favorable targets for mineral exploration. Figure 6 shows a 12-digit WBD that identifies a finer resolution geographic area that corresponds to the geochemically anomalous stream sediment sample data used to select the 8-digit WBD shown in figure 5.

Watersheds that were mapped as part of the Rowan and others (1984) study were georectified and compared in GIS with 12-digit WBDs and watersheds delineated using the GIS Weasel. This comparison was made to help understand the issues of scale when doing detailed geochemical investigations such as was done for the Rowan and others (1984) study and regional mineral assessments involving much larger areas. Twelve-digit WBDs and watersheds delineated using the GIS Weasel are shown overlain on watersheds delineated in the field (figs. 7 and 8). These comparisons clearly show that the field-derived watersheds provide a much higher spatial resolution for more precisely identifying geochemically anomalous subwatershed areas.

Sediment Provenance

GIS Tools Available for Sediment Provenance

Thematic mapping of point data using proportional symbols, though useful in identifying samples and specific sites with anomalous element concentrations, does not account for adjacent, larger areas that could contribute to and influence the geochemistry at a specific point location. The NHD and WBD datasets are useful in identifying larger areas that potentially contribute water and sediment to a site.

In order to identify potential NHD streams that contribute sediment to a sample site using GIS analysis, a point must intersect a NHD stream segment. Identification of WBD contributing areas, especially 8- and 12-digit HUCs, are not as dependent on accurate sample locations due to the large areas

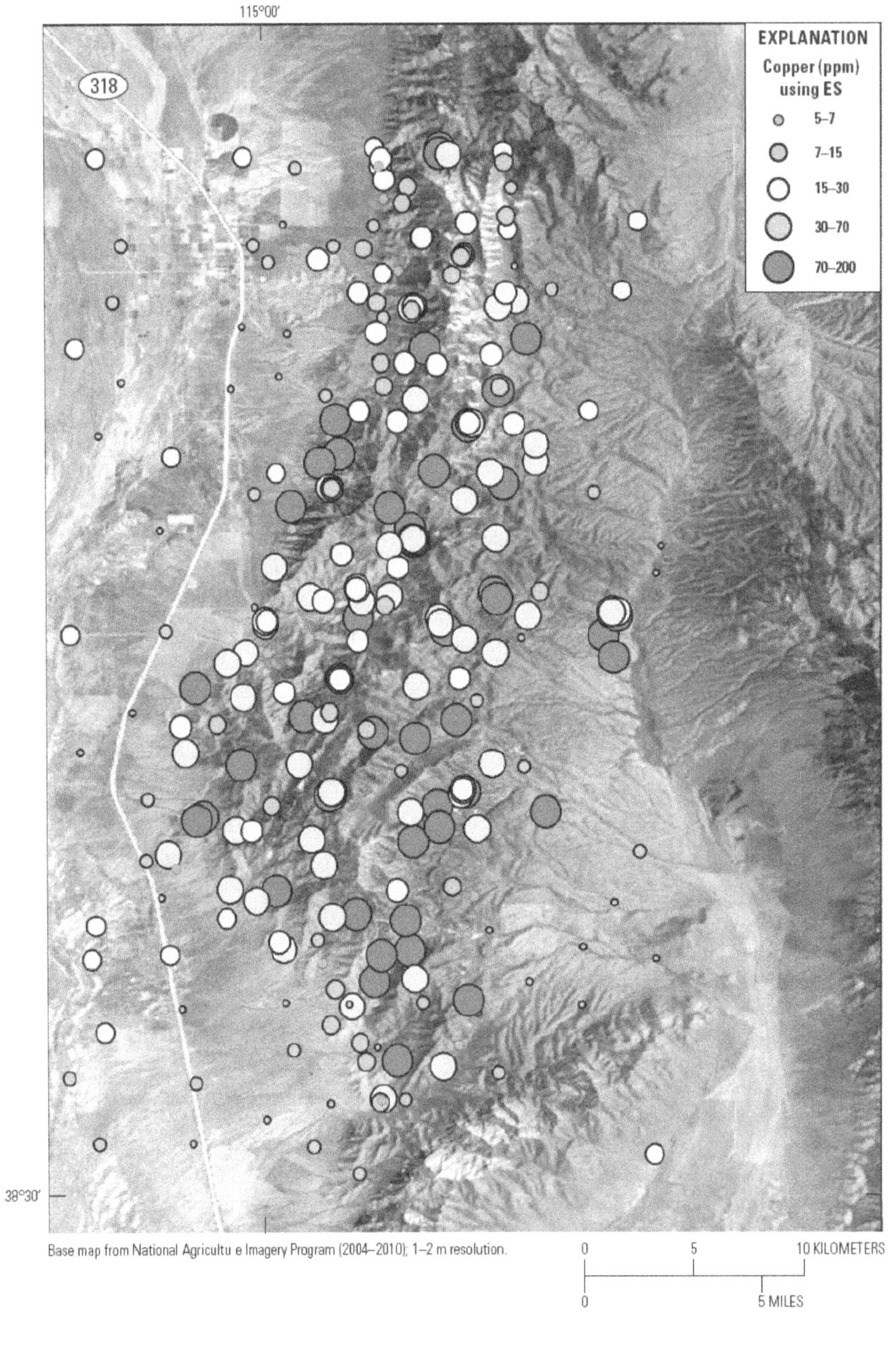

Figure 3.   Copper (Cu) concentrations (in parts per million, [ppm]) determined for sediment sample sites as part of study by Rowan and others (1984). Relative Cu concentration increases proportionally with increase in symbol size. Warmer colors (yellow, orange, red) represent higher element concentrations. Average sample spacing is 757 m. ES, semi-quantitative emission spectrography.

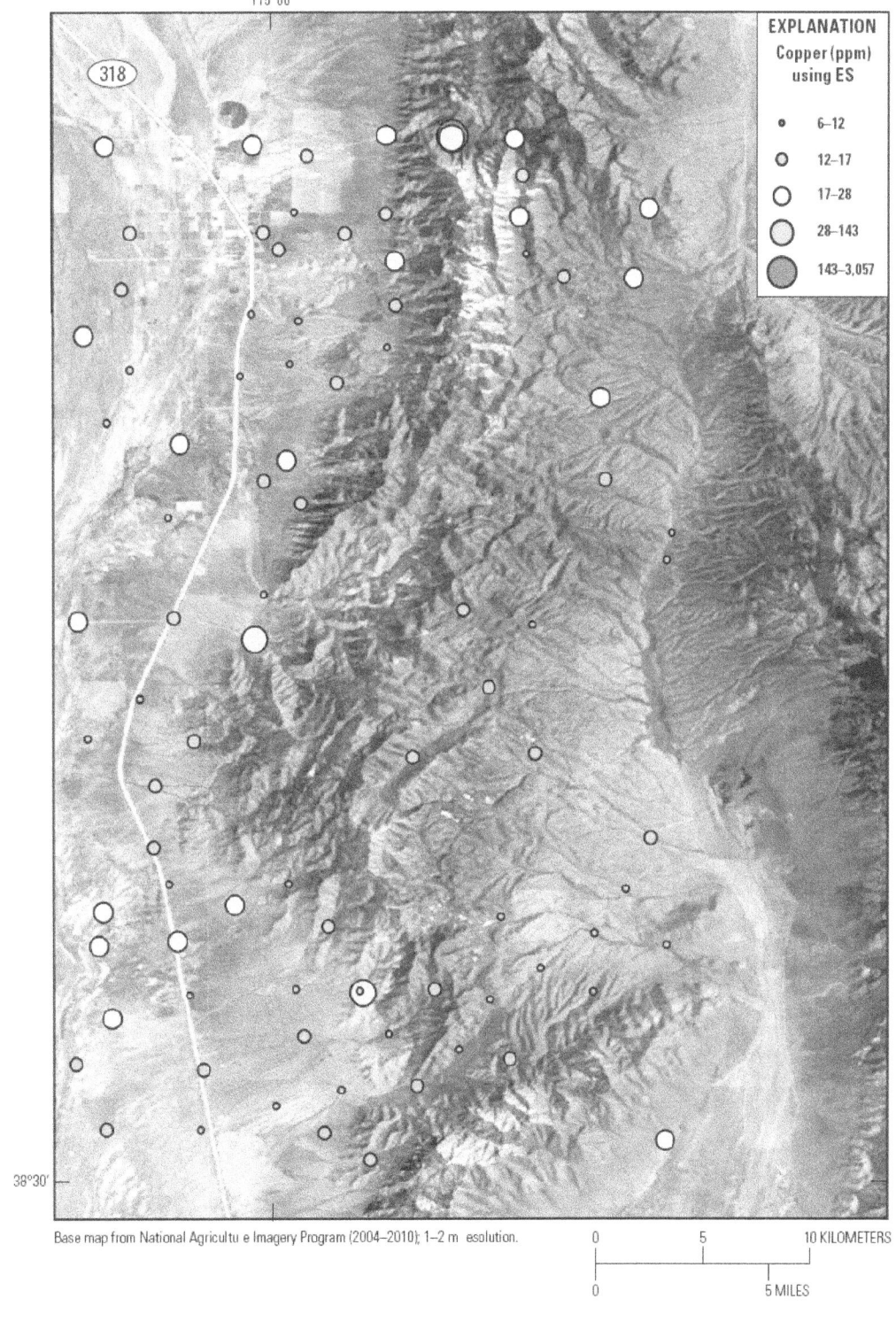

Figure 4. Copper concentrations (in parts per million, [ppm]) for National Uranium Resource Evaluation (NURE) sediment samples in the Egan Wilderness Study Area. Symbol size is proportional to increasing element concentration. Warm colors have higher concentration and cool colors represent lower concentration. Average sample spacing is 2,145 m. ES, semi-quantitative emission spectrography.

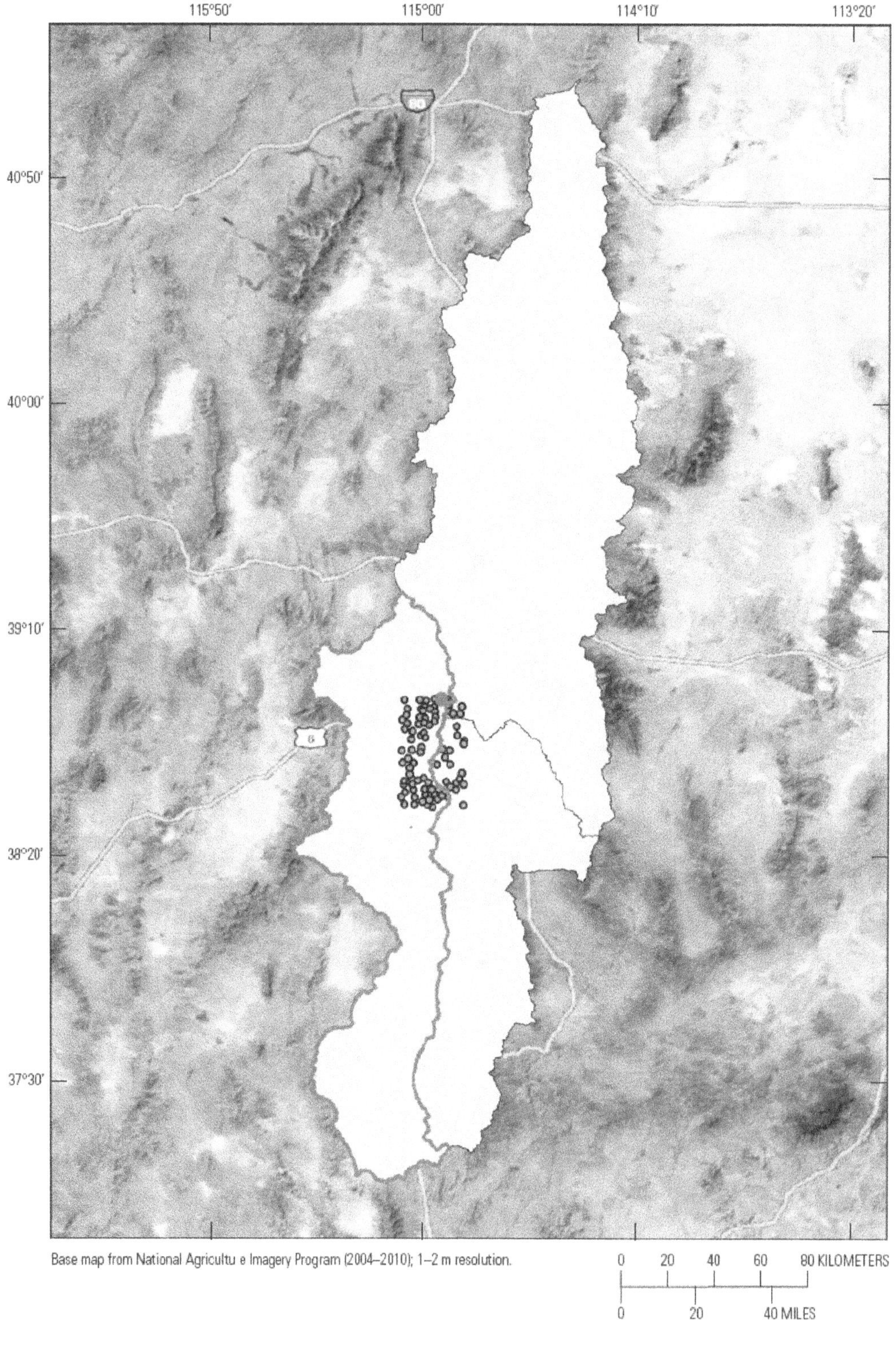

Figure 5.   Three, 8-digit Watershed Boundaries (WBDs) encompassing study area. National Uranium Resource Evaluation (NURE) sediment samples (blue and red dots) are in the Bureau of Land Management Egan Wilderness Study Area. A geographic information system select-by-location query identifies WBD (red outline) corresponding to stream sediment sample (red dot) having an anomalous copper concentration.

115°00'

39°00'

39°30'

Base map from National Agricultu e Imagery Program (2004–2010); 1–2 m esolution.

0   5   10   15   20 KILOMETERS

0   5   10 MILES

Figure 6.   Twelve-digit Watershed Boundary Dataset (WBD) for study area. Christmas Tree Canyon WBD (red outline) was selected using ArcGIS 10.1 and select-by-location query, where selected WBD corresponds to National Uranium Resource Evaluation (NURE) sediment sample (red dot) having relatively high copper concentration. Blue dots represent NURE sediment sample sites having relatively lower copper concentrations.

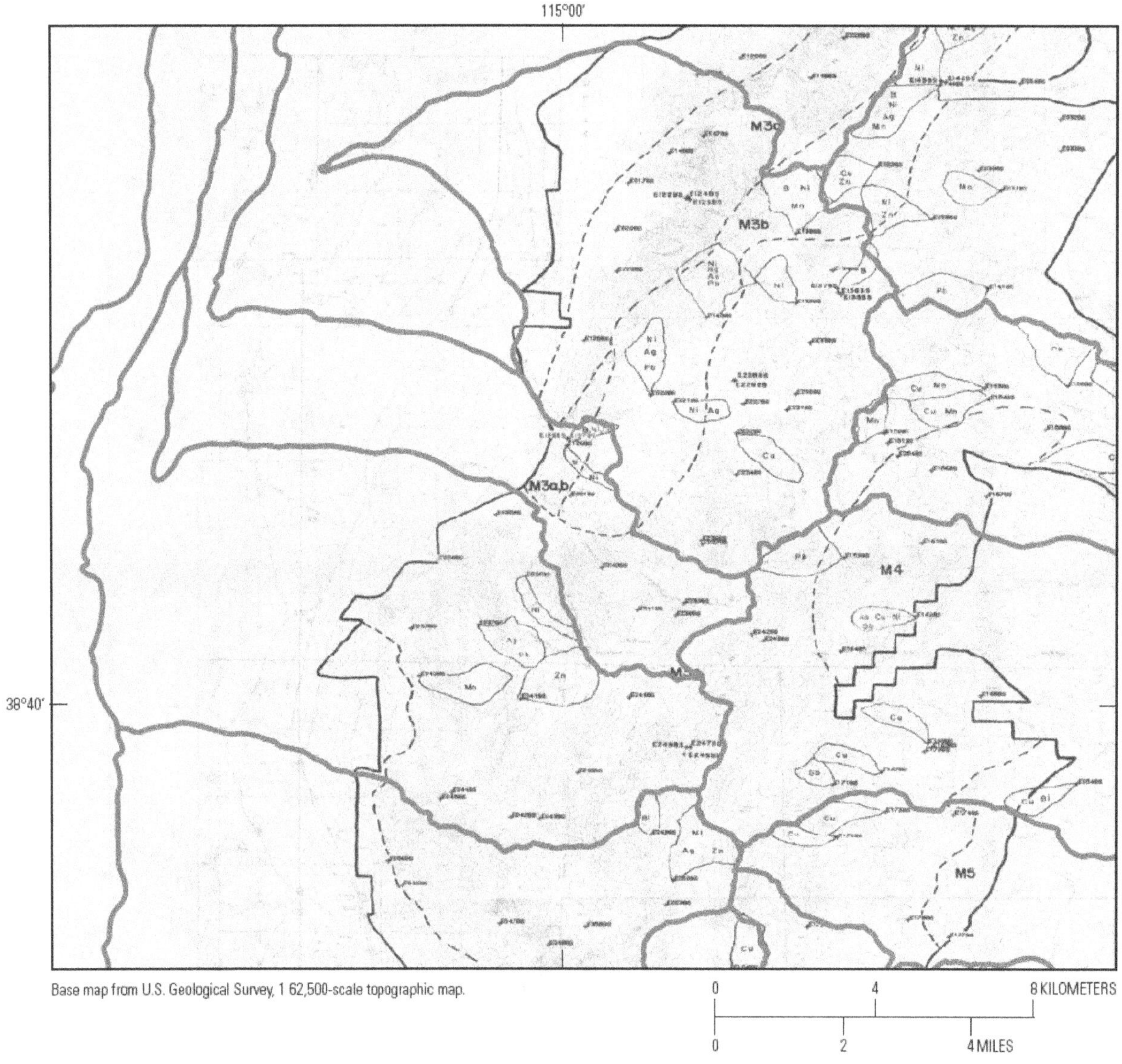

Figure 7.   Twelve-digit Watershed Boundary Dataset (red outlines) and georeferenced watersheds from Rowan and others (1984).

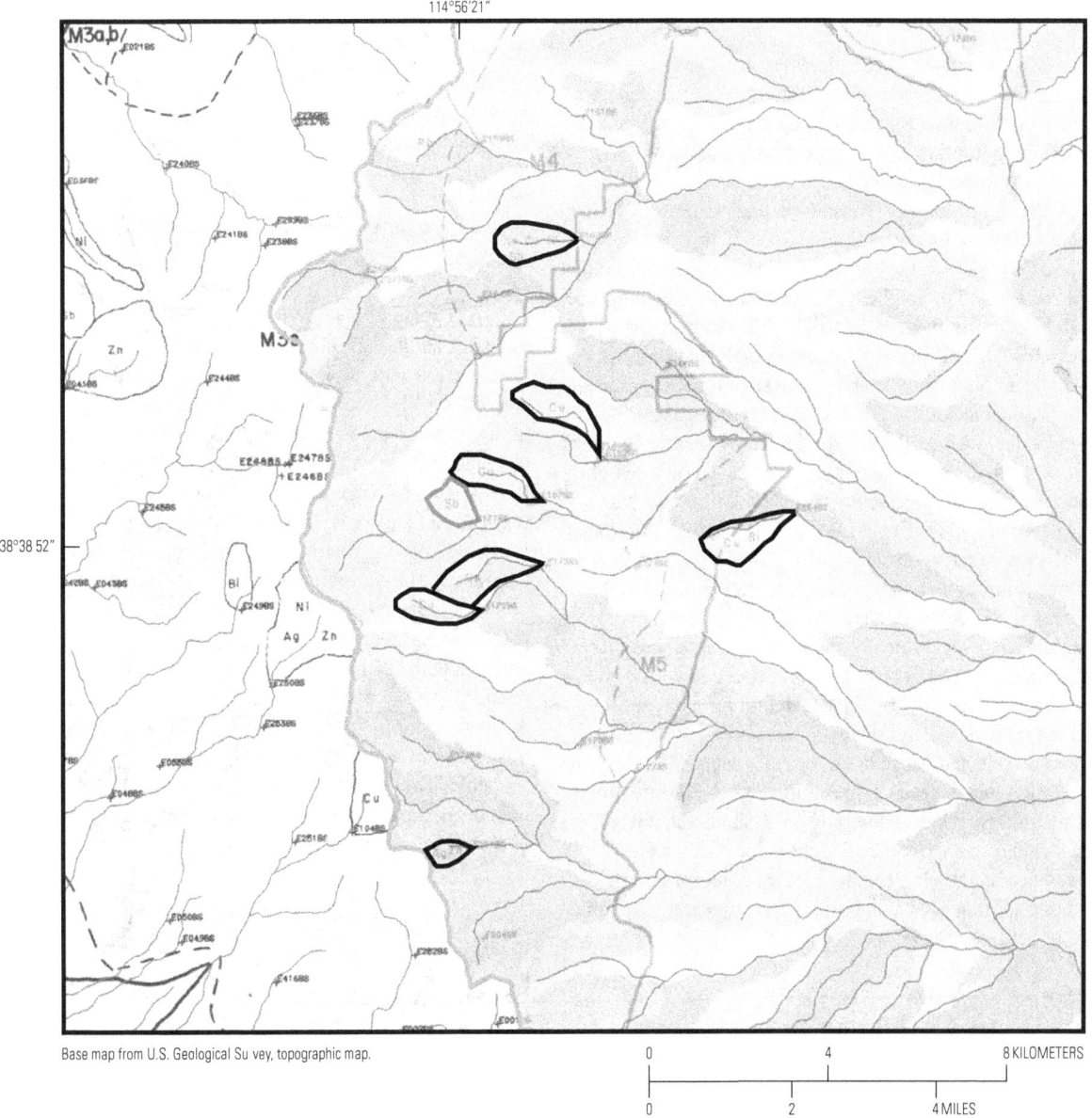

Figure 8.   Two-plane watersheds delineated using the GIS Weasel (pastel polygons) and georeferenced watersheds from Rowan and others (1984). Watersheds delineated by Rowan and others (1984) are within heavy black outlines and have a bisecting National Hydrography Dataset (NHD) stream; watershed outlined in red lacks NHD stream.

involved in GIS overlay analysis or select-by-location queries. A watershed can be identified as a potential contributing area to a site simply when a sample plots within a WBD.

Linear Referencing in ArcGIS

To assess whether geochemical data used in this analysis intersect the NHD network, the raw sample data were plotted with the NHD streams. An intersect by location query in Arc-GIS 10.1, in addition to visual inspection, revealed that few to no sediment samples intersect the NHD network. Utilizing the ArcGIS 10.1 proximity (Near) tool, the locations of NURE sediment sample points were found to plot within an average distance of 90 m of NHD streams. In addition, distribution of samples was found to parallel NHD stream channels suggesting that the samples were collected on tributary streams.

ArcGIS 10.1 has linear referencing tools that enable points to be automatically snapped to a stream network based on search-tolerance distance rules. This involves using the "Locate Features Along Routes" tool to create a table of points and locations that are snapped to a stream based on a search-tolerance distance. Subsequently, the "Make Route Event Layer" is used to add the events to the GIS analysis environment for subsequent analysis.

Once samples were snapped to a stream and an event layer created, the definition query tool in ArcGIS 10.1 was used to identify geochemically anomalous sample sites that subsequently could be geographically associated with NHD streams and watersheds (fig. 9). The definition query tool is part of the layer properties dialogue box and allows a subset of data to be selected based on query rules such as an element concentration that exceeds a threshold value. NHD stream routes that intersect sites exceeding a threshold value were then selected using a select-by-location query in ArcGIS. NHD segments that intersect the selected points are highlighted and subsequently can be output as a separate shapefile for use in guiding mineral exploration.

Hydrography Event Management Tools

The Hydrography Event Management (HEM) tools are a utility add-on to ArcGIS 10.1 that is used to create point and area features that can be spatially linked via GIS to NHD data. HEM tools provide additional automated functionality for snapping point-geochemical features to NHD stream features. The automated functionality and efficiency in snapping point features to streams is especially important for analysis of multiple geochemically anomalous sample sites such as is required for a national mineral resource assessment. Steps to build HEM features are described in online documentation prepared by Bureau of Land Management (2011) available at *http://usgs-mrs.cr.usgs.gov/hemtool/HEMToolManual_v24/ HEMToolManual_v24.html.*

Creating HEM point events involves making a new Arc-GIS geodatabase and a feature class that is a placeholder for storing imported point features (Bureau of Land Management, 2011). Point features are imported to the new geodatabase using the HEM "Import to Events" tool. A search distance is set during import that will be used to snap points to a NHD flowline network. The initial results of the snapped points are stored in a quality control geodatabase, and a HEM dialogue allows each snapped point feature to be reviewed for accuracy. All points that are accepted and verified to be accurate are uploaded to a geodatabase as final HEM point events. A GIS "relate" or "join" permits linking HEM point features that are associated with a NHD network to point data having geochemical attributes that can be selected to establish anomalous samples.

A "Create Points to Flags or Barriers" tool in HEM provides for additional functionality not readily available with the linear referencing tools in ArcGIS 10.1. Once point events are created in HEM and snapped to the NHD using distance criteria and an associated NHD network, the HEM Create Points to Flags or Barriers tool permits flags to be globally created from selected events. The flags serve as locations on a NHD network to then apply ArcGIS 10.1 utility-network analysis functions, such as finding the upstream contributing area from anomalous stream sediment sample sites (fig. 10).

The selected NHD streams from the steps described above can subsequently be output as a shapefile for additional analysis with WBD data. This is done by first "returning results as a selection" from the "Utility Network Analyst, Analysis Options" dropdown dialogue, solving again for upstream accumulation area, and exporting selected data as a shapefile. The exported shapefile of anomalous streams can then be associated with WBDs using either NHD reach codes, or a select-by-location query in ArcGIS (fig. 11).

Geologic Attributes of Anomalous National Hydrography Dataset Streams and Watershed Boundary Datasets

Once anomalous NHD streams and WBDs are identified, other geoscience datasets can be analyzed to determine whether specific geologic units are geographically associated with these hydrologic features. This study determined summary geologic-units area statistics for 12-digit WBDs having anomalous Cu concentrations shown in figure 11 that intersected the geologic units mapped by Crafford (2007) (fig. 12, table 2). These types of summary information can be statistically analyzed using regression analyses such as those described by Yager and others (2010) to determine correlations between a geologic-unit type area and element concentration of sediment or water collected at a watershed outlet. Summary statistics of WBD geology, once compiled using GIS, can also

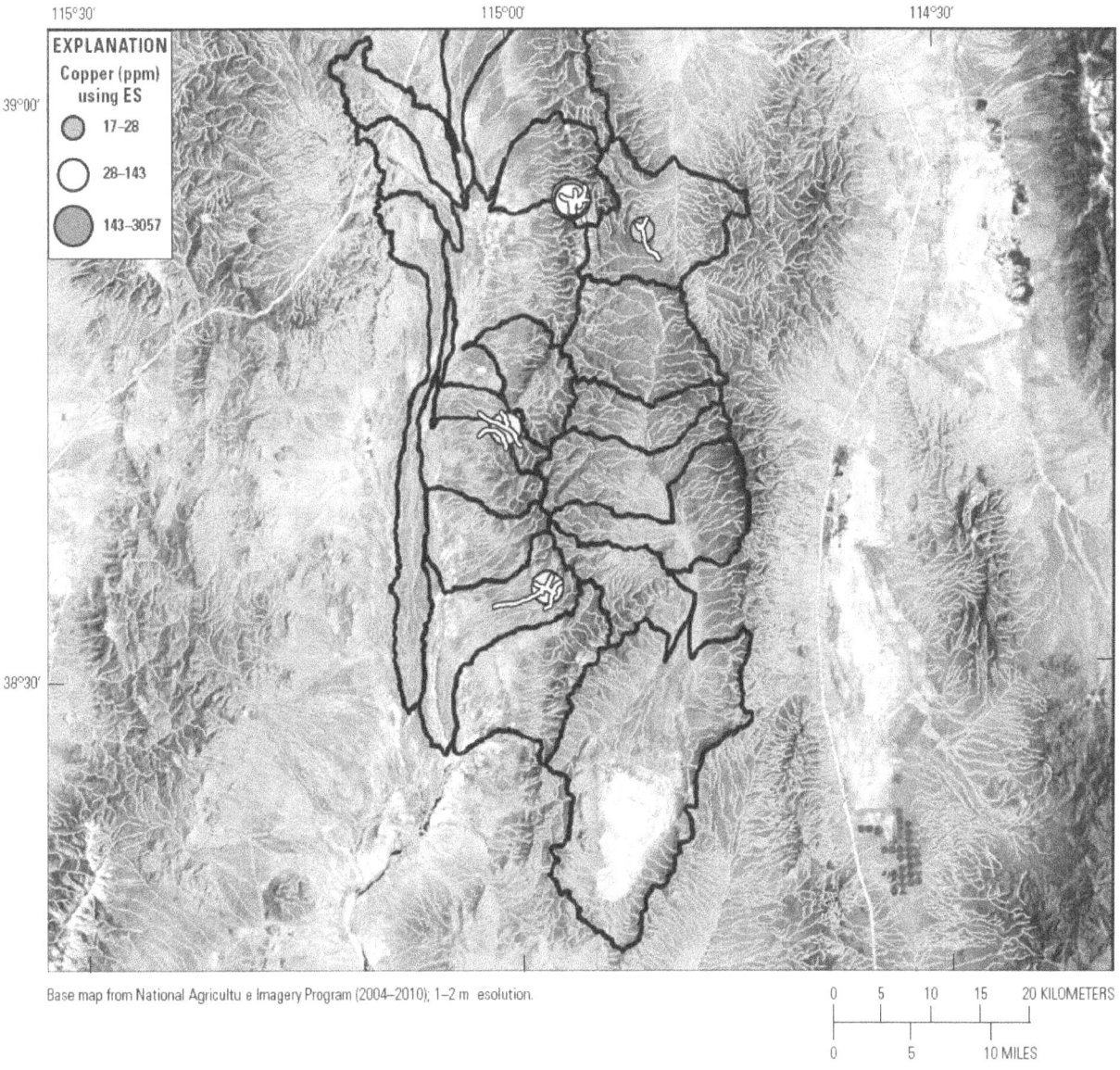

Figure 9. Copper (Cu) concentrations (in parts per million [ppm]) for National Uranium Resource Evaluation (NURE) sediment samples. A definition query in ArcGIS was used to select Cu concentrations that exceeded an element threshold value. Larger symbols are proportional to increasing element concentration. ArcGIS 10.1 linear referencing tools were used to create point events that are associated with National Hydrography Dataset (NHD) streams. A select-by-location query permitted NHD stream segments that intersect the point locations to be highlighted (yellow lines) and output as a separate shapefile that could be used to help guide mineral assessments. ES, semi-quantitative emission spectrography.

Table 2.   Summary statistics for geologic units determined by analyzing the geographic information system intersection of Egan Wilderness Study Area 12-digit watershed boundaries that have copper concentrations exceeding 20 parts per million.

[Geologic unit abbreviations taken from geographic information system database attribute field "FMATN" that is part of the Nevada geologic coverage of Crafford (2007) and available online at *http //pubs.usgs.gov/ds/2007/249/*; WBD, Watershed Boundary Dataset]

WBD and geologic units	Unit area (m²)	Unit area (%)	WBD and geologic units	Unit area (m²)	Unit area (%)
White Horse Pasture-White River			Christmas Tree Canyon		
Dg	12,820,848	19	Єu	12,177,493	7
Dse	2,988,335	4	Dg	34,703,153	20
Dsi	2,403,235	4	Ds	46,237	9
Dsu	1,343,348	2	MD	149,528	28
Mc	5,741,571	9	Ol	38,195,842	23
MDp	793,294.5	1	SOu	20,087,766	12
Mj	709,771.4	1	Tov	1,333,026	1
Ml	5,205,813	8			
Msw	2,065,696	3			
Oe	1,870,733	3			
Oes	1,742,581	3			
Op	2,767,119	4			
PPl	23,292,613	35			
Sl	2,848,305	4			
Middle Cattle Camp Wash			Head of Nine Mile Spring		
Dg	36,248,624	14	Dg	3,283,103	7
Ds	15,936,116	6	Dse	1,955,054	4
MD	48,452,056	19	MD	18,765,373	42
Ol	3,507,066	1	Msc	2,341,109	5
Par	43,353,272	17	Oe	1,378,887	3
PIP	51,944,917	21	Oes	992,943	2
Sou	20,087,766	8	Ol	2,864,754	6
Tot	11,525,899	5	Op	2,767,119	6
Tov	19,004,344	8	PIP	6,626,559	15
			Sl	1,436,269	3
			Sou	424,943	1
			Tos	2,048,154	5
			Dg	3,283,103	7
Silver Creek-Cave Valley Wash			Silver Springs-White River		
Єl	28,196,144	14	Dg	2,953,882	4
Єm	5,095,352	3	Mc	5,741,571	9
Єu	9,608,389	5	Ml	2,107,078	3
Dg	33,742,980	17	Msw	1,903,293	3
Ds	15,493,723	8	PPl	23,292,613	35
MD	18,664,469	9	PPs	7,399,545	11
Ol	19,043,728	10	Tkvu	18,001,513	27
PIP	9,492,479	5	Tsp	477,149	1
Tos	20,231,631	10	Tvt	3,923,981	6
Tot	27,397,288	14			
Tov	586,0745	3			
Tys	4,728,128	2			

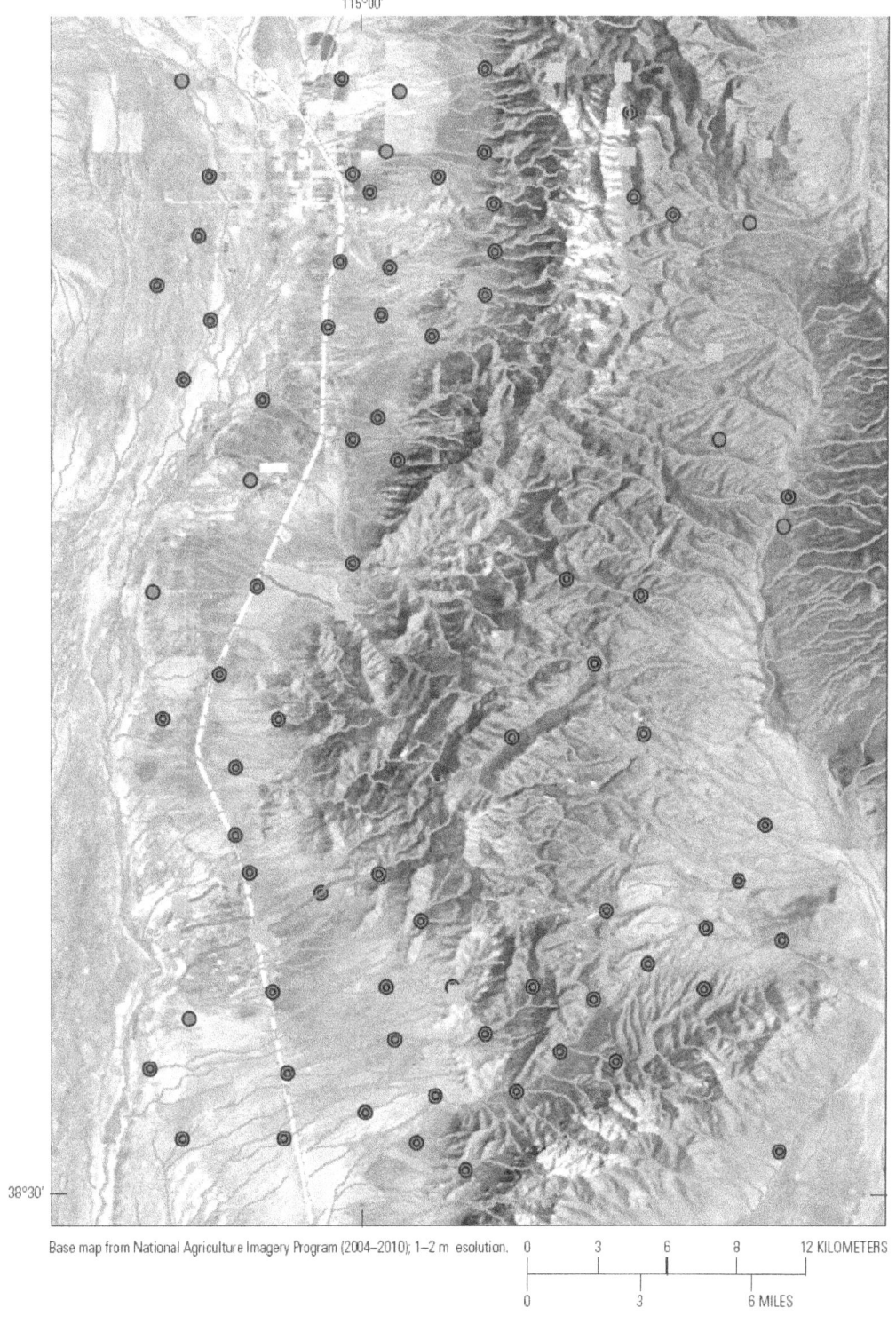

Base map from National Agriculture Imagery Program (2004–2010); 1–2 m esolution.

Figure 10. Results of using the Hydrography Event Management (HEM) tools to identify potential upstream contributing areas of sediment where copper (Cu) exceeds a geochemical threshold value (20 parts per million [ppm] in this example). National Hydrography Dataset (NHD) stream network shown in blue; all National Uranium Resource Evaluation (NURE) sediment sample points in red. HEM tool "flags" (green squares) created from HEM point events where Cu exceeds a 20-ppm threshold value. ArcGIS Utility Network Analyst tools were used to trace upstream contributing areas (red lines) relative to flagged locations. Points were automatically snapped (using the HEM tools) to the NHD network based on a specified distance criteria. Points that were not snapped to streams are represented by single points; nested points are locations snapped to streams.

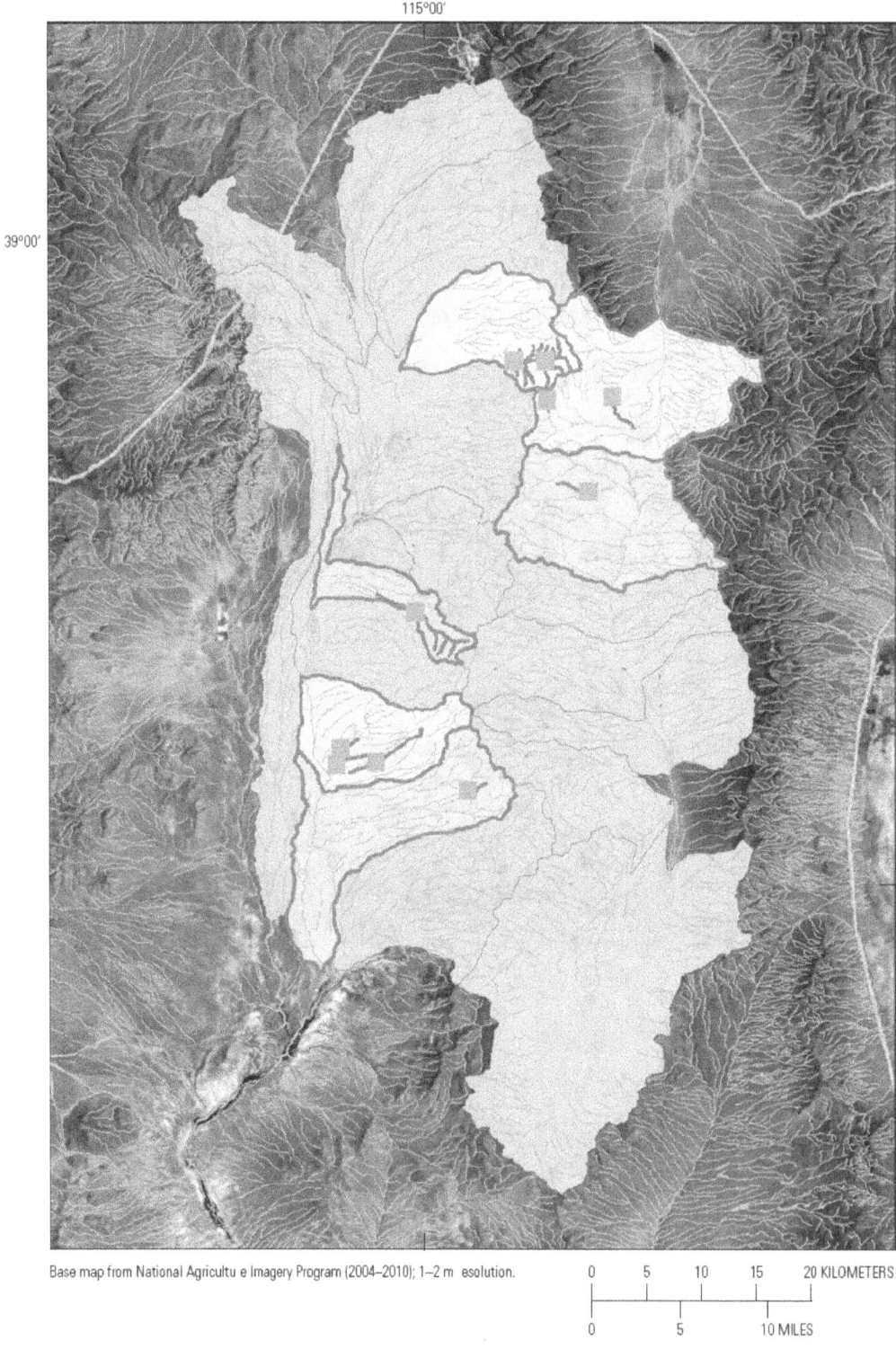

115°00'

39°00'

Base map from National Agricultu e Imagery Program (2004–2010); 1–2 m esolution.

0   5   10   15   20 KILOMETERS

0   5   10 MILES

Figure 11.   Twelve-digit hydrologic-unit code boundaries (Watershed Boundary Dataset) in red that intersect National Hydrography Dataset (NHD) stream segments (red line segments) determined using the Hydrography Event Management (HEM) tools where copper exceeds a threshold value of 20 parts per million. NHD stream network in blue. HEM tool "flags" (green squares) created from HEM point events.

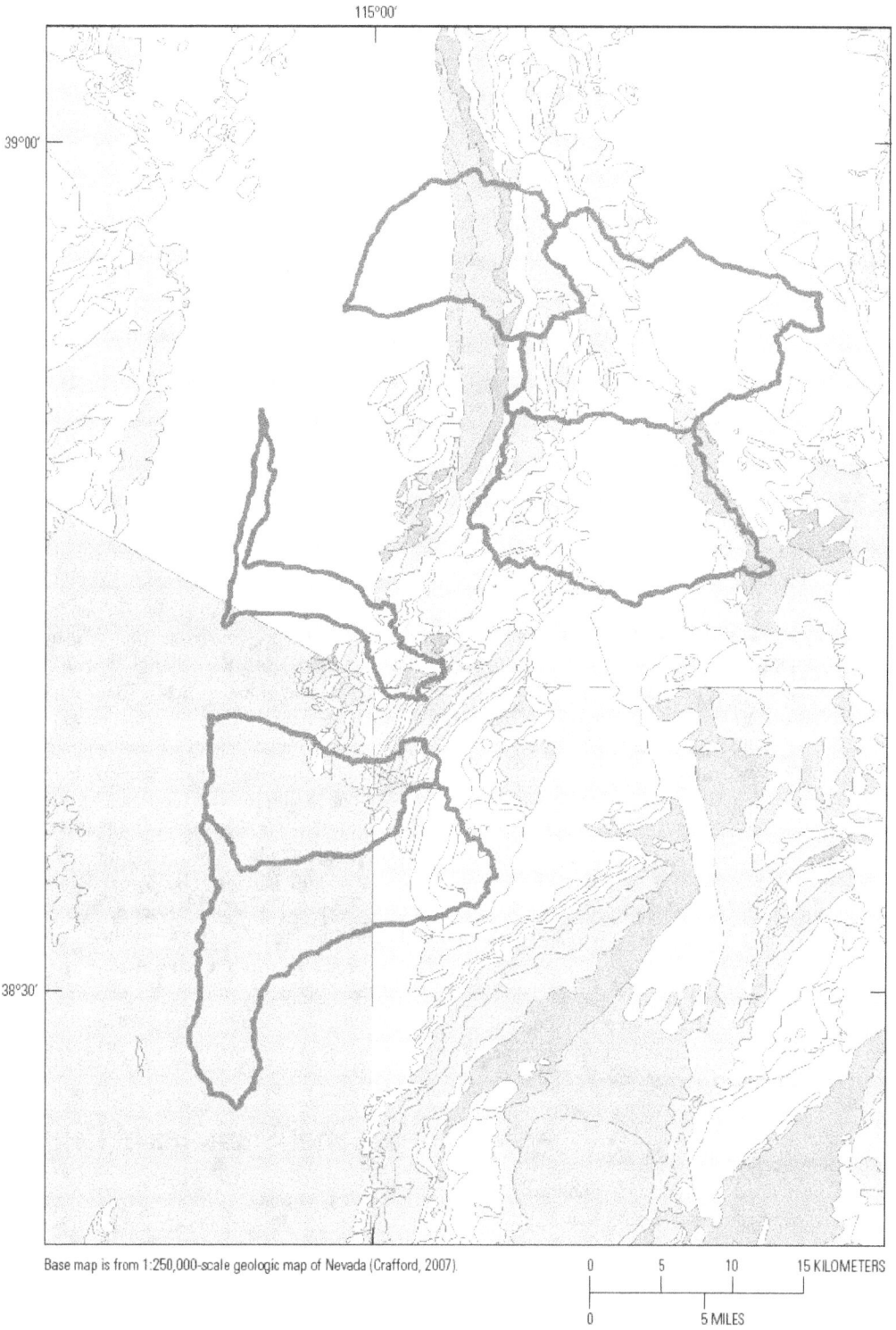

115°00'

39°00'

38°30'

Base map is from 1:250,000-scale geologic map of Nevada (Crafford, 2007).

0   5   10   15 KILOMETERS

0   5 MILES

Figure 12.   Twelve-digit hydrologic-unit codes (Watershed Boundary Dataset [WBD] boundaries in red) discussed in figure 6 overlain on geology from Crafford (2007). ArcGIS 10.1 intersection tools were used to compile geologic unit summary statistics of WBD polygons shown in table 2.

be visualized in association with other map data. This type of GIS display helps to elucidate trends in geology and associated WBD geochemistry (fig. 13).

NHD streams that are identified as upstream contributing areas of sediment can also be analyzed in GIS to determine the geographic intersection of geologic units and geochemically anomalous streams. Subsequent statistical analysis either in GIS or statistical software could be used to test correlations between geologic units intersected by geochemically anomalous streams.

Alluvial Fans

In the desert southwest and in many areas of the Western United States, alluvial fans have formed at watershed outlets at the base of mountain ranges. Alluvial fans are especially prevalent in the Basin and Range physiographic province where mountain ranges have a high potential energy for hydrologic weathering. The material weathered from the mountain ranges is deposited as alluvial fans where there is an abrupt change to low gradients in the adjacent basins. Relative ages of adjacent fans are important to consider because the source of stream sediments sampled could be from multiple stream channels whose source could be from separate watersheds. The relative ages of fans can also be complex. An example of the geographic relation of multiple fans that have formed in the BLM wilderness study area is shown in figure 14A.

In understanding the stream sediment provenance of alluvial fans, a combination of GIS datasets and processing and analysis steps is proposed to help delineate possible sediment-contributing areas to the fans. This type of effort is time intensive and would likely only be feasible in selected areas as part of a national mineral resource assessment. In the desert southwest, high-resolution (1-m) imagery such as that available from National Agriculture Imagery Program (NAIP) enables alluvial fans to be identified at the edge of mountain ranges. The plan geometry of alluvial fans is easily discerned using the imagery and permits these features to be digitized as polygon features. Once digitized, the alluvial fans could be thought of as inverted watersheds with the fans representing the eroded core of the watersheds in which they were derived.

There are multiple methods in GIS to geographically associate alluvial fans with hydrologic features that could aid in establishing sediment provenance. Polygon identifiers can be assigned based on the NHD hydrologic features that a fan intersects. For example, NHD reach codes or watershed boundary identifiers could be attached to the alluvial fan polygon-attribute table using GIS intersection tools. Another approach is to use the Utility Network Analyst tool to set a "flag" (a point location on a hydrologic network) and solve for the "Find Upstream Accumulation" along the NHD network (fig. 14B). The selected accumulation area subsequently could be geographically associated with digitized alluvial fans, highlighting the possible sources of sediment delineated from

the NHD. This is a GIS methodology requiring a significant investment in time because flags would need to be established using the Utility Network Analyst tool that would identify the highest elevation point on the fan.

A more automated approach could be used that would add GIS database attributes of NHD and WBD hydrologic features to geographically associated alluvial fans. The spatial overlay tool "Spatial Join" attaches attributes from one feature to another based on a geographic relation. When a geographic intersection of features exists between the target features and the join features, such as fans and NHD features, the joined attributes from the join features are written to the output feature class. A similar overlay tool, "Intersect," would determine the geometric intersection of the input alluvial fans and hydrologic identity features that the fans overlap. Attributes of the identity features would be attached to the alluvial fan polygon features that would then allow their selection by query in GIS. This approach could aid in automating selection of multiple fans that contain anomalous stream sediment geochemistry. However, the NHD reach codes and WBD identifications that are associated with high metal concentrations would first need to be delineated. Once reach codes that have high metal concentrations are established, alluvial fan polygons with the same reach codes previously determined using GIS overlay could be selected using a standard GIS query.

The proximity tools in ArcGIS 10.1 could also be effective in determining the possible source areas of geochemical anomalies on alluvial fans. If a geochemically anomalous sample is identified on a fan, anomalous sediment and rock samples that are in proximity could be identified based on search radius criteria. Because dilution effects would likely cause downstream element concentrations for sediments deposited on alluvial fans to be lower compared to element concentrations in a potential source area, a selection query might only involve those samples with element concentrations equal to or greater than the fan geochemical sample. Potential source areas for geochemical anomalies in sediments would be selected based on upstream rock samples having higher element concentrations than those observed on the fan.

Geospatial Statistics of Point Data

A data clustering, hot-spot-analysis tool in ArcGIS 10.1 was also used to analyze NURE and BLM wilderness sediment data. Hot-spot analysis can be used to evaluate the concentration of a geochemical element at multiple, adjacent sites (Getis and Ord, 1992). A hot spot exists when a site and its geographic neighbors have a high element concentration. A Z score (test of statistical significance based on standard deviation) is calculated for each data point. Clusters of data having statistically significant positive Z scores are an indication of a geochemical anomaly. Low Z scores indicate a clustering of low values that could be used to identify the geochemical baseline for an element and areas that can be excluded as mineral exploration targets.

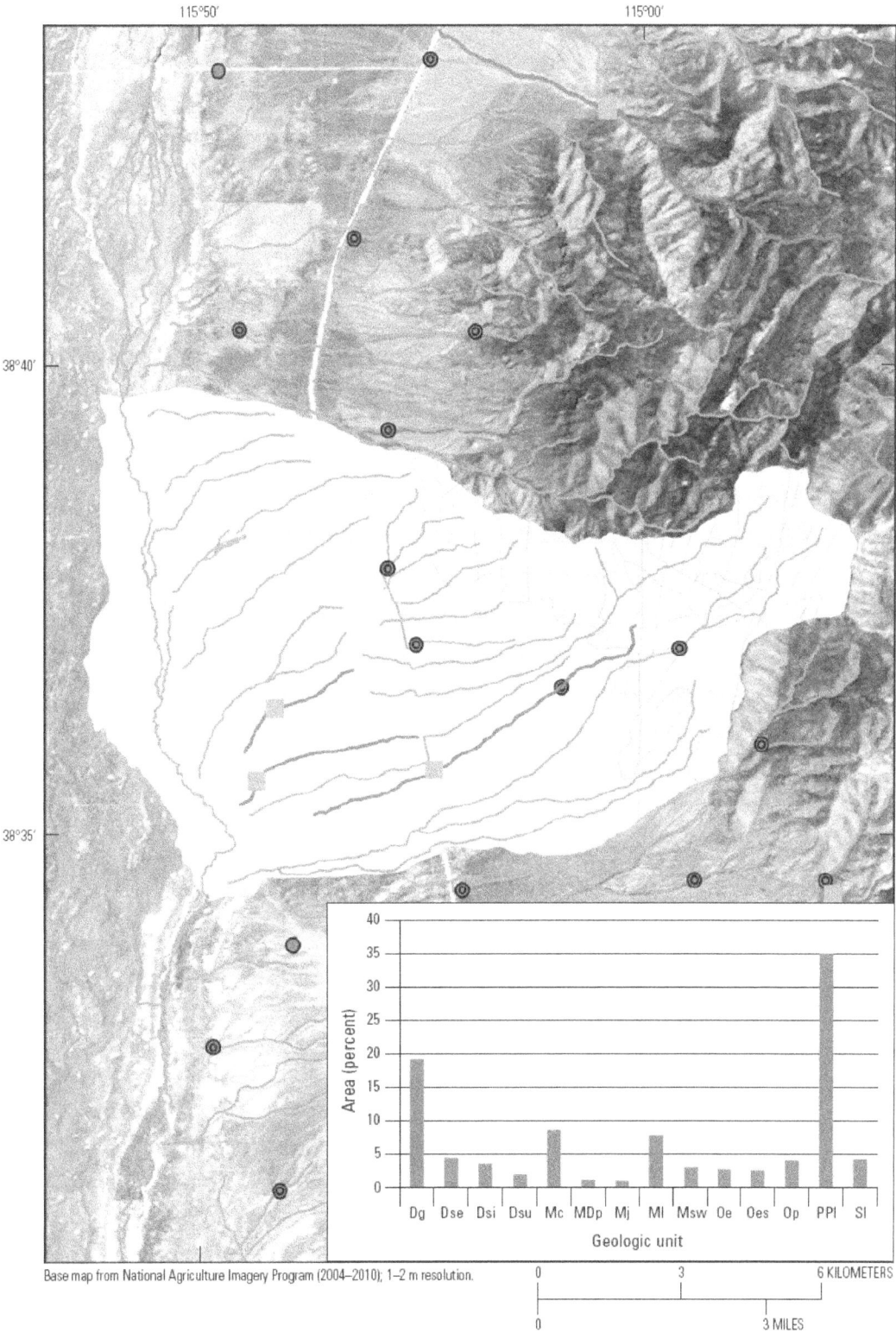

Figure 13.   Area statistics for geologic units determined from GIS intersection of geologic map of Crafford (2007) that intersect the White Horse Pasture-White River 12-digit Watershed Boundary Dataset. Watershed was found to have possible anomalous copper concentrations as shown in figure 11. Histogram (inset) shows proportion of bedrock units in the watershed. Symbology same as shown in figure 10, which identifies upstream contributing areas of potentially geochemically anomalous National Uranium Resource Evaluation (NURE) stream sediment sample sites.

Hot-spot analysis of Cu concentrations in sediments for a BLM wilderness area by Rowan and others (1984) and NURE studies yielded different results. Different 12-digit WBDs were identified as having high Z scores for Cu (figs. 15 and 16). The different results could be a function of the higher sampling density used by Rowan and others (1984). Hot-spot analysis of Cu analyzed in rocks as part of the Rowan and others (1984) study, however, identified the Christmas Tree Canyon 12-digit

WBD as anomalous (fig. 17), which is consistent with similar hot-spot analyses of Cu in NURE stream sediments. Relatively high Cu concentrations were also identified in the Christmas Tree Canyon 12-digit WBD for all rock analyses culled from the NGDB. These results indicate that integration of multiple datasets would be useful in helping delineate geochemical anomalies for mineral resource investigations (fig. 18).

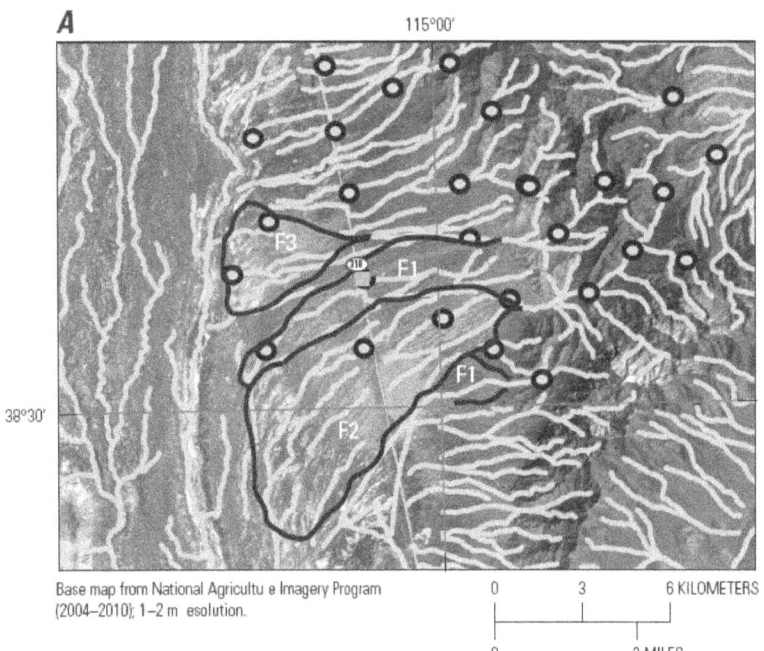

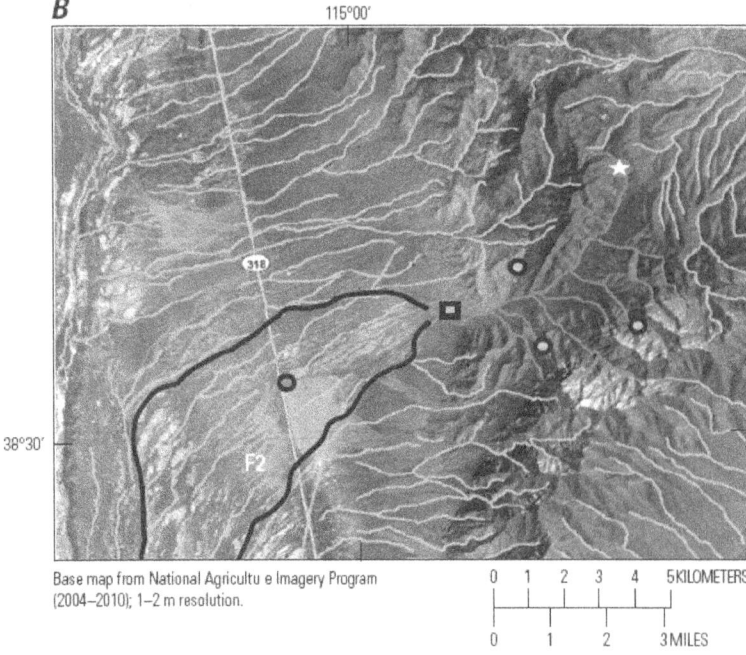

Figure 14.   (*A*) Example of alluvial fans in basins adjacent to mountain ranges (east part of image). National Hydrography Dataset (NHD) streams in blue. Multiple generations of fans are interpreted to have formed and are listed in order of oldest (F1) to youngest (F3). (*B*) Upstream accumulation area of NHD streams (red) was determined using the set flag (green square) option in the ArcGIS 10.1 Utility Network Analyst. Star symbol represents farthest upstream sample that is potentially contributing sediment to alluvial fan (F2).

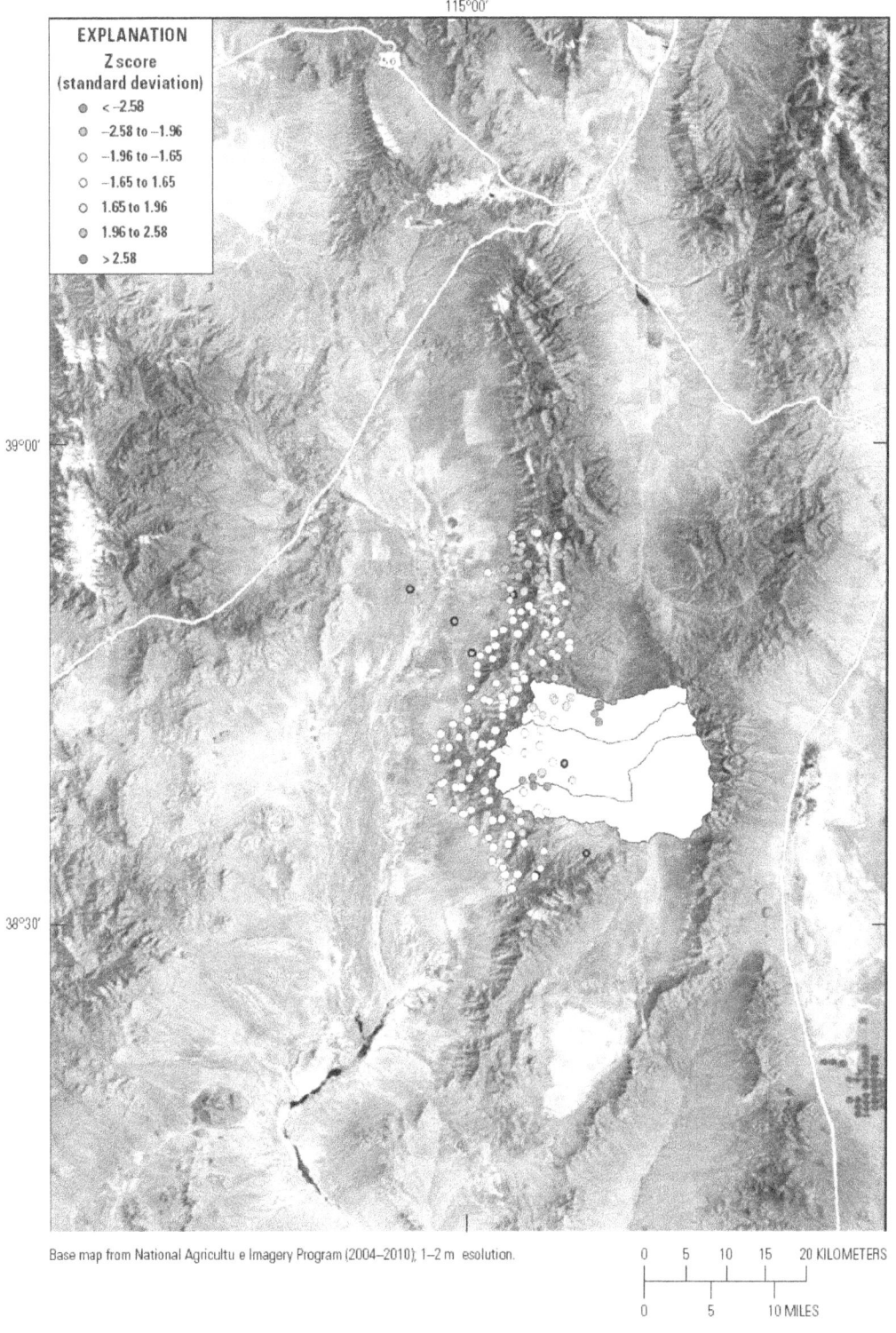

Figure 15.   Hot-spot analysis of copper concentrations (in parts per million) determined for sediment sample sites as part of Rowan and others (1984) study. An ArcGIS select-by-location query was used to identify the 12-digit Watershed Boundary Datasets symbolized as solid polygons outlined with black lines that correspond with Z scores greater than or equal to 2.58 standard deviations (red dots).

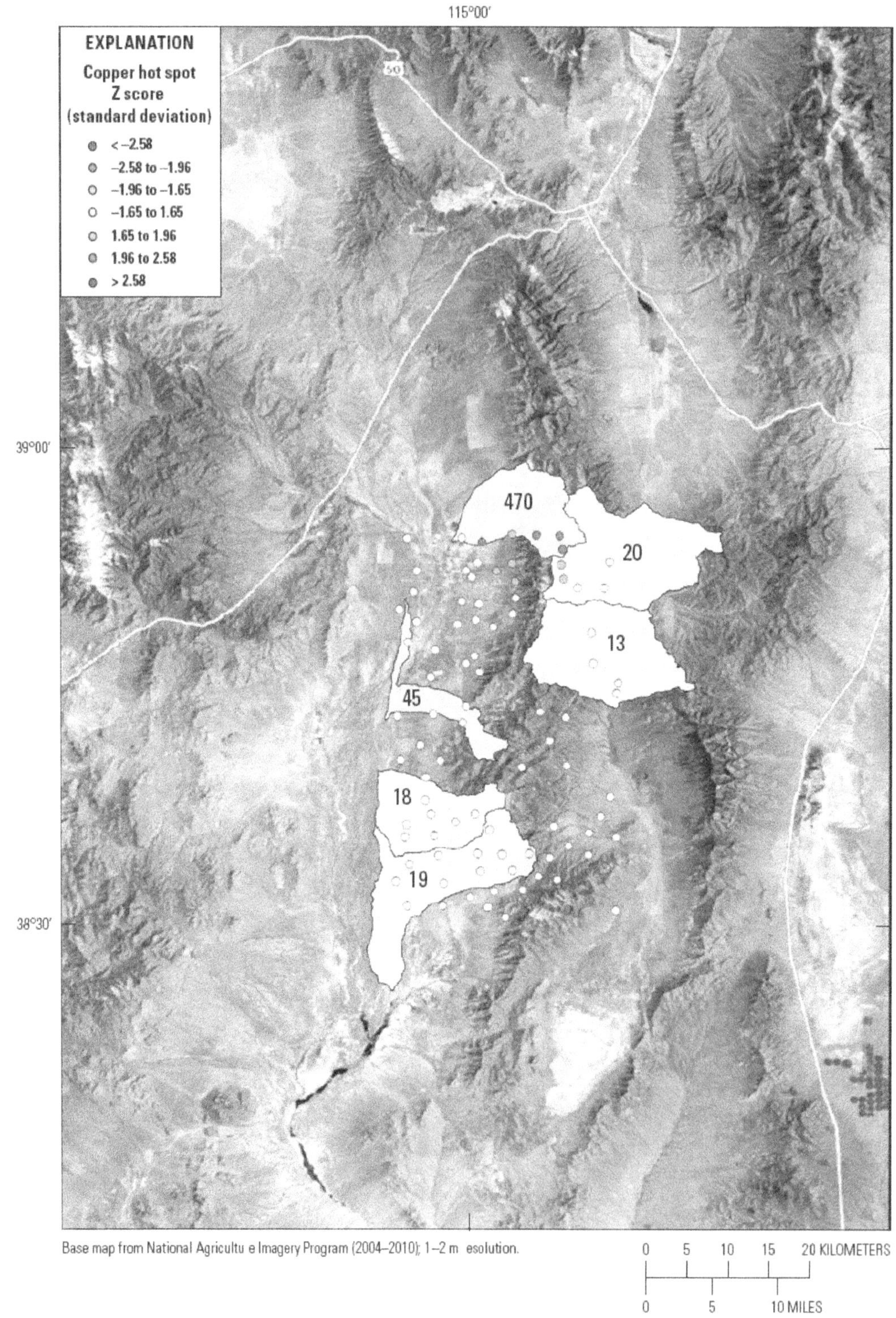

Figure 16.   Results of ArcGIS 10.1 hot-spot analysis. National Uranium Resource Evaluation (NURE) sediment copper (Cu) concentrations are represented in standard deviation above and below the mean. Anomalously high Cu concentrations (red dots) coincide with Christmas Tree Canyon, 12-digit Watershed Boundary Dataset (WBD) (pink area labeled 470). Mean Cu concentrations labeled for each WBD.

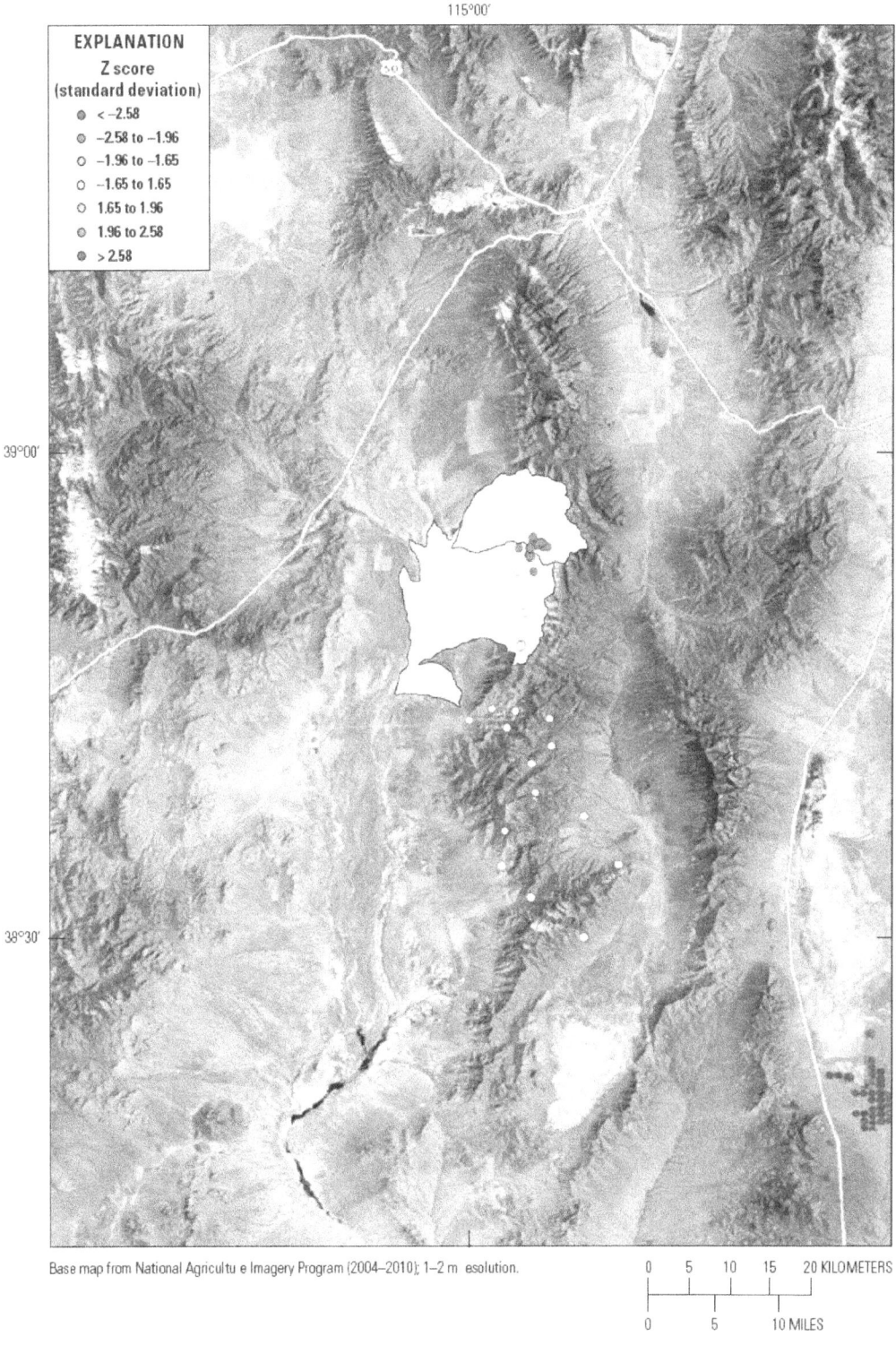

Figure 17. Hot-spot analysis of copper (Cu) concentrations (in parts per million) determined for rock samples as part of Rowan and others (1984) study. An ArcGIS select-by-location query was used to identify the 12-digit Watershed Boundary Datasets (WBDs), symbolized as solid polygons outlined with black lines that correspond with Z scores greater than or equal to 2.58 standard deviations. The northernmost WBD area (Christmas Tree Canyon) was also identified as having high Z scores for National Uranium Resource Evaluation (NURE) sediment Cu hot-spot analysis.

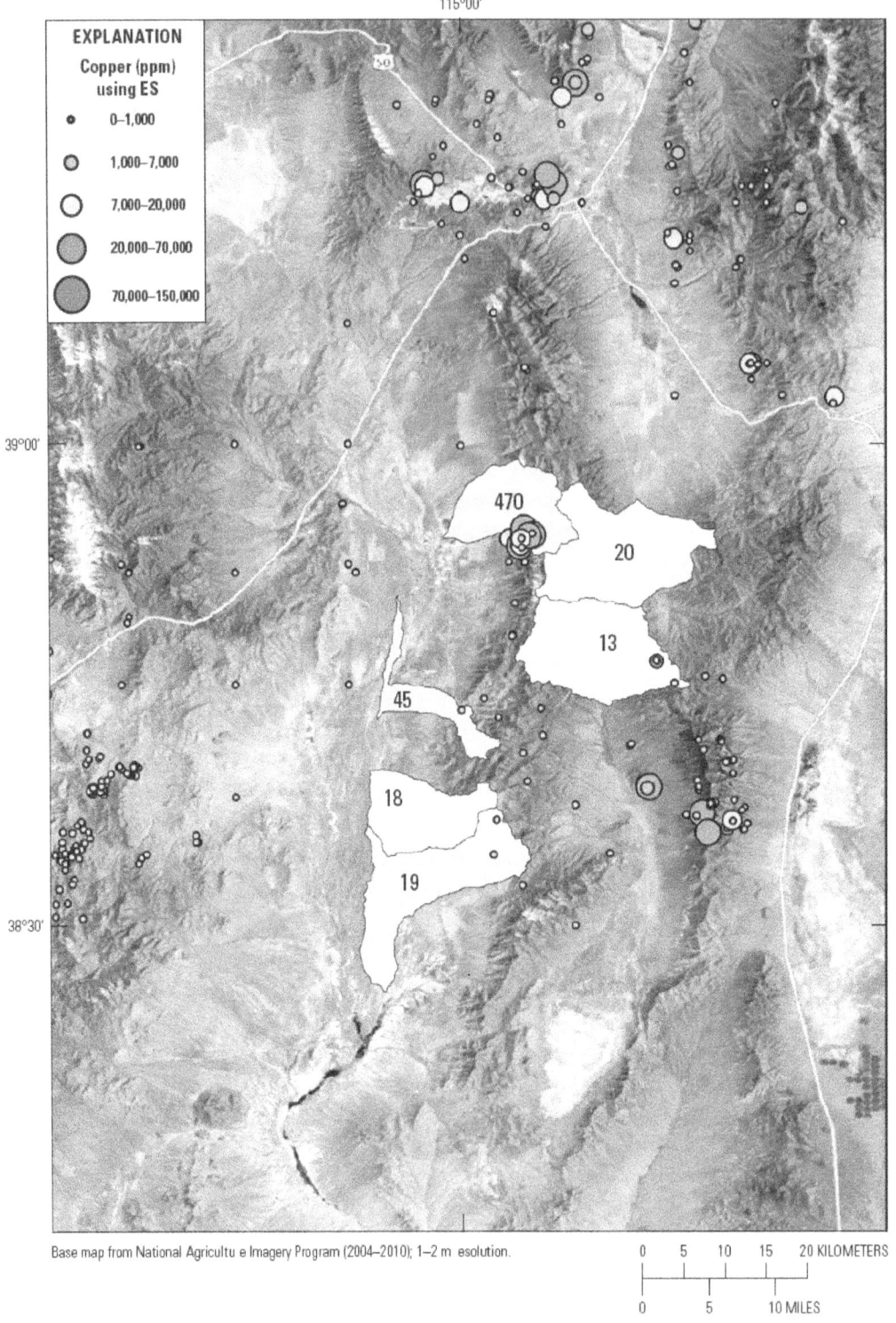

Figure 18.   Copper (Cu) concentrations (in parts per million [ppm]) for rocks that were culled from the National Geochemical Database. Increasing symbol size is proportional to higher Cu concentrations determined by emission spectrographies (ES). Warmer colors (red, yellow) represent relatively high concentrations; lower concentrations are represented by cool colors (blue, green). Mean Cu concentrations labeled for each Watershed Boundary Dataset. Note that high Cu concentrations for rocks correspond with high average Cu concentrations in sediments especially for the Christmas Tree Canyon watershed (pink, labeled 470).

Discussion

GIS methodologies outlined in this report are an efficient means to analyze large areas as part of a mineral assessment. GIS tools available in ArcGIS that were not readily available in previous USGS mineral assessments are effective in quickly analyzing data having x-y and geochemical attribute information to highlight geochemical anomalies at point localities. Anomalous point localities can subsequently be analyzed using select-by-location queries to identify geographically associated WBDs. The precision of potentially anomalous areas identified is dependent on the scale of watersheds used for a select-by-location query.

Hydrologic datasets such as the NHD that were specifically designed to function with ArcGIS analysis tools, for example, linear referencing, HEM, and Utility Network Analyst, are useful in delineating stream sediment provenance. Attributes that are coded in the NHD stream network allow the upstream and downstream relations to a sample site to be identified. This functionality could be a large improvement in more accurately identifying the precise source locations of metal anomalies.

Analysis of legacy geochemical data involving BLM wilderness, NURE, and NGDB datasets sometimes yielded differing results. A higher sampling density used as part of the Rowan and others (1984) study compared to the NURE study could partly explain differences, especially in the hot-spot analysis (figs. 15 and 16). In contrast, rock data in the study by Rowan and others (1984) showed a Cu anomaly in a similar location as the NURE stream sediment data (fig. 17). This suggests that analysis of geochemical constituents determined for all available sample media types would be required in mineral assessments to more accurately identify geochemically anomalous areas.

Conclusions

This report discusses an effective GIS methodology to efficiently analyze legacy data as part of a USGS national mineral resource assessment. Tools are available in ArcGIS 10.1 to analyze point data having x-y and geochemical attribute information that can be used as a fundamental first step in delineating geochemical anomalies. The NHD and associated WBD data provide a geographic context in which to identify areas encompassing and adjacent to geochemically anomalous points. The NHD has GIS attributes to identify upstream and downstream relations relative to sample points that can be used to determine stream sediment and water provenance.

Determining the provenance of geochemically anomalous sediments is essential in identifying the areas to focus mineral resource investigations in the field. The HEM tools are useful in analyzing point data in a NHD hydrologic network context to efficiently evaluate stream sediment sample provenance. Geochemically anomalous upstream contributing areas are identified with the aid of HEM and Utility Network Analyst tools that could be useful to guide mineral exploration in the field. Mineral resource investigations could also benefit from detailed investigations in selected areas utilizing such tools as GIS Weasel that defines detailed WBDs and hydrologic networks that can be geographically associated with USGS legacy data.

Acknowledgments

The authors wish to thank William J. Carswell, Jr., and Ariel Doumbouya for discussions regarding GIS functionality of the NHD datasets and the HEM tools. Reviews by John Horton and Ariel Doumbouya are greatly appreciated. This work was funded by the USGS Mineral Resources Program and the Central Mineral and Environmental Resources Science Center.

References Cited

Brady, L.M., Gray, Floyd, Wissler, C.A., and Guertin, D.P., 2001, Spatial variability of sediment erosion processes using GIS analysis within watersheds in a historically mined region, Patagonia Mountains, Arizona: U.S. Geological Survey Open-File Report 01–267, 48 p.

Bureau of Land Management, 2011, Hydrography event management tools—User guide, ver. 2.4: Bureau of Land Management Website, available at *http://usgs-mrs.cr.usgs.gov/hemtool/HEMToolManual_v24/HEMToolManual_v24.html.*

Carranza, E.J.M., 2004, Usefulness of stream order to detect stream sediment geochemical anomalies: Geochemistry: Exploration, Environment, Analysis, v. 4, p. 341–352.

Carranza, E.J.M., and Hale, Martin, 1997, A catchment basin approach to the analysis of reconnaissance geochemical-geological data from Albay Province, Philippines: Journal of Geochemical Exploration, v. 60, no. 2, p. 157–171.

Church, S.E., Fey, D.L., and Unruh, D.M., 2008, Trace elements and lead isotopes in modern streambed and terrace sediment—Determination of current and premining geochemical baselines, chap. E12 *of* Church, S.E., von Guerard, Paul, and Finger, S.E., eds., Integrated investigations of environmental effects of historical mining in the Animas River watershed, San Juan County, Colorado: U.S. Geological Survey Professional Paper 1651, p. 575–642.

Crafford, A.E.J., 2007, Geologic map of Nevada, *with a section on* A digital conodont database of Nevada, by A.G. Harris and A.E.J. Crafford: U.S. Geological Survey Data Series, 249, 46 p., scale 1:250,000, 1-CD (version 1.1 was released in 2008).

Gesch, Dean, Oimoen, Michael, Greenlee, Susan, Nelson, Charles, Steuck, Michael, and Tyler, Dean, 2002, The National Elevation Dataset: Photogrammetric Engineering and Remote Sensing, v. 68, no. 1, p. 5–33.

Getis, Arthur, and Ord, J.K., 1992, The analysis of spatial association by use of distance statistics: Geographic Analysis, v. 24, no. 3, p. 189–206.

Granitto, Matthew, Yager, D.B., and Hofstra, A.H., 2005, Geochemical data for the Great Basin [abs.]: A subset of the USGS new national geochemical database: Geological Society of America Abstracts With Programs, v. 37, no. 7, p. 380.

PRISM Climate Group, 2002, PRISM dataset: Oregon State University Website, *http://prism.oregonstate.edu/*.

Rowan, L.R., Hofstra, A.H., and Gordon, W.D., 1984, Reconnaissance geochemical assessment of metallic mineral resource potential South Egan Range Wilderness Study Area (NV 040-168), White Pine, Lincoln, and Nye Counties, NV: U.S. Geological Survey Open-File Report 84–782, 61 p.

Seoane, J.C.S., and De Barros Silva, Ardemirio, 1999, Gold-anomalous catchment basins: A GIS prioritization model considering drainage sinuosity: Journal of Geochemical Exploration, v. 67, p. 335–344.

Simley, J.D., and Carswell, W.J., Jr., 2009, The national map—Hydrography: U.S. Geological Survey Fact Sheet 2009–3054, 4 p.

Smith, S.M., 1997, National geochemical database: Reformatted data from the National Uranium Resource Evaluation (NURE) Hydrogeochemical and Stream Sediment Reconnaissance (HSSR) Program, ver. 1.40 (2006): U.S. Geological Survey Open-File Report 97–492, accessed Feb. 1, 2006 at *http://pubs.usgs.gov/of/1997/ofr-97-0492/index.html*.

U.S. Geological Survey, 2002, Assessment of undiscovered deposits of gold, silver, copper, lead, and zinc in the United States–A portable document (PDF) recompilation of USGS Open-File Report 96–96 and Circular 1178: U.S. Geological Survey Open-File Report 02–198, 2719 p.

U.S. Geological Survey and U.S. Department of Agriculture, Natural Resources Conservation Service, 2011, Federal standards and procedures for the national Watershed Boundary Dataset (WBD), 2d ed.: U.S. Geological Survey Techniques and Methods 11–A3, 62 p.

Viger, R.J., and Leavesley, G.H., 2006, The GIS Weasel user's manual: U.S. Geological Survey Techniques and Methods, Book 6, Chap. B4, 201 p., available at http://wwwbrr.cr.usgs.gov/projects/SW_MoWS/pubs/viger_pubs/viger-Pubs.shtml.

Viger, R.J., 2008, The GIS Weasel: An interface for the development of geographic information used in environmental simulation modeling: Computers and Geosciences, v. 34, p. 891–901.

Yager, D.B., Burchell, Alison, and Johnson, R.H, 2010, A geologic and anthropogenic journey from the Precambrian to the new energy economy through the San Juan volcanic field, *in* Morgan, L.A., and Quane, S.L., eds., Through the generations: Geologic and anthropogenic field excursions in the Rocky Mountains from modern to ancient: Geological Society of America Field Guide 18, p. 193–237.

Yager, D.B., and Folger, H.W., 2003, Maps showing concentrations of 13 elements from stream sediments and soils throughout the Humboldt River basin and surrounding areas, northern Nevada: U.S. Geological Survey Miscellaneous Field Studies Map MF–2407–A–M, 1:500,000 scale.

www.ingramcontent.com/pod-product-compliance
Lightning Source LLC
Chambersburg PA
CBHW081413170526

45166CB00010B/3321